BIRDING
~ at the ~

BRIDGE

In Search of Every Bird
on the Brooklyn Waterfront

Heather Wolf
Foreword by David Lindo

THE EXPERIMENT

NEW YORK

BIRDING AT THE BRIDGE: *In Search of Every Bird on the Brooklyn Waterfront*
Text and photos copyright © 2016 by Heather Wolf
Map copyright © 2016 by Holly Graham

The Experiment, LLC
220 East 23rd Street, Suite 301
New York, NY 10010-4674
www.theexperimentpublishing.com

The Experiment's books are available at special discounts when purchased in bulk for premiums and sales promotions as well as for fundraising or educational use. For details, contact us at info@theexperimentpublishing.com. Many of the designations used by manufacturers and sellers to distinguish their products are claimed as trademarks. Where those designations appear in this book and The Experiment was aware of a trademark claim, the designations have been capitalized.

Library of Congress Cataloging-in-Publication Data

Names: Wolf, Heather.
Title: Birding at the bridge : in search of every bird on the Brooklyn
  waterfront / Heather Wolf ; foreword by David Lindo.
Description: New York : The Experiment, [2016] | Includes bibliographical
  references and index.
Identifiers: LCCN 2015048252 (print) | LCCN 2016010479 (ebook) | ISBN
  9781615193134 (pbk.) | ISBN 9781615193141 (ebook)
Subjects: LCSH: Bird watching--New York (State)--New York. | Birds--New York
  (State)--New York--Pictorial works. | Birds--Counting--New York
  (State)--New York. | Brooklyn (New York, N.Y.)
Classification: LCC QL684.N7 W65 2016 (print) | LCC QL684.N7 (ebook) | DDC
  598.072/347471--dc23
LC record available at http://lccn.loc.gov/2015048252

ISBN 978-1-61519-313-4
Ebook ISBN 978-1-61519-314-1

Cover and text design by Sarah Smith

Manufactured in China
Distributed by Workman Publishing Company, Inc.
Distributed simultaneously in Canada by Thomas Allen and Son Ltd.
First printing June 2016

10 9 8 7 6 5 4 3 2 1

To Bob and Lucy Duncan

~

Ring-billed gull, Pier 1

Contents

Foreword

This book tells the story of one woman's quest to find as many bird species as possible on her local patch, but it's also much more: a story of discovery, a fascinating journey, a testament to the wonders all around us, an inspiration.

I like to say that whatever city I'm in, I spend so much time gazing skyward at birds that it looks as if I'm staring up at the Manhattan skyline for the first time. So it seemed too perfect that, when I met Heather Wolf in the fall of 2015 for a birding stroll around Brooklyn Bridge Park, our day of memorable sightings came against the stunning backdrop of lower Manhattan. With Heather as my guide, New York has never looked better—nor has it ever been so fun.

Heather is an extraordinary woman, a life force with abundant energy, and I sensed her excitement each time we rounded a path's corner, where it seemed we'd inevitably discover yet another interesting species lurking in a bush or foraging on the parkland floor. Heather's enthusiasm was infectious, and her love for birds, birding, and, above all, her local patch comes across just as clearly on the page. I know from experience what it's like to be with her when she spots a bird; for those who haven't had the pleasure, I can say that *Birding at the Bridge* vividly captures the thrill.

Heather's philosophy is one that I totally subscribe to: Urban areas are filled with natural wonder and are worthy places to explore, watch, and study. When she moved near Brooklyn Bridge Park—still being built at the time—she didn't see a construction site, nor did she see just another urban park; she saw a new and much-needed wildlife habitat. Now, with her remarkable photos and thousands of sightings, she's proved that her own neighborhood park holds riches that stand up to the more famous New York birding sites, like Jamaica Bay Wildlife Refuge and the avian mecca that is Central Park. Thanks to her, Brooklyn Bridge Park is now very much a landmark on the ornithological map.

Heather's story somewhat echoes mine. I, too, discovered an unlikely local patch, Wormwood Scrubs in west London. I was ridiculed for even contemplating birding there. Yet, 150 species and many rarities later, it is now a go-to site for many a passing birder visiting London. I can see the same thing happening on the Brooklyn waterfront.

Above all, this book proves that living in a city is no reason not to be a birder. As Heather amply illustrates through her words and photography, cities hold their own natural treasures that are no less beautiful or breathtaking for their urban environs. Just grab a birding book and a pair of binoculars, find a local patch, and bring an open mind. You will soon be finding birds that you never realized were there. Just remember: Anything can turn up anywhere at anytime.

David Lindo
The Urban Birder
November 10, 2015

*Manhattan skyline view
from Vale Lawn*

The Quest

've never been a morning person. I'm still not a morning person. You might be wondering how someone like me got into birding, a pastime that often requires waking up at the crack of dawn. Birds are the only thing, barring an emergency, that makes me rise before the sun. Once I developed this passion, the possibility of spotting an interesting species or observing a new behavior became an obsession. I said good-bye to the snooze button. I braved freezing temperatures. I flew to an unfamiliar location in Texas, rented a car, and set out on a two-hour drive at 3:30 AM to get somewhere else I'd never been by sunrise. I was pulled over for speeding at 27 mph (in a 25 mph zone) and had to explain to a police officer that I was on my way to see a scaled quail.

Birding, also known as bird-watching, is sometimes perceived as humdrum, a hobby reserved for retirees and senior citizens. Nothing could be further from the truth. How could observing the closest living descendants of dinosaurs be boring? I've traveled with the circus and cruised the Caribbean as a lecturer, but none of that can match the excitement I've experienced while birding. To me, it's exhilarating, challenging, saddening, maddening, and addictive. There's potential for adventure at every moment, as interesting birds exhibiting entertaining behavior can show up nearly anywhere. The search for birds becomes never-ending. Once you've become a birder, your spouse will be shouting at you to watch the road as you scan power lines for birds while driving. When you visit your family, they'll be baffled why you're renting a car and heading out to some desolate location in their state before dawn. Welcome to birding.

I lived in Brooklyn from 2001 to 2006, working as a software developer in midtown Manhattan, a block from the Empire

State Building. I didn't watch birds. I loved New York City, but eventually I felt the need to escape the fast-paced lifestyle, at least for a while. So in 2006, I moved to Florida's Pensacola Beach, a beautiful twelve-square-mile stretch of sugar-white sand on the Gulf of Mexico. Being away from big-city distractions made it much easier to notice birds. At first I loved watching shorebirds dig in the soaked sand for small mollusks, then scurry up the beach as the next wave rolled in. Their streamlined flight just inches above the sea impressed me. I watched them, I admired them, but I wasn't a birder—yet.

My earliest recollection of becoming interested enough in birds to start studying them came after an adventurous encounter on the beach. As I walked on a path bordered by a wide expanse of sand dunes, taking in the salt air, I heard a high-pitched whistling. Then I noticed a bird flying toward me. Actually, it was flying straight at me. It looked a bit like a small gull, but with thinner wings and a more pointed yellow beak. There I was, alone, with nothing but the sand, sun, water, and this beautiful yet ominous bird that started to dive-bomb me! Both thrilled and scared, I had to do something to spare my eyeballs, so I ran up the path and out of the danger zone. I continued my walk, knowing we would meet again on my way back home. When I approached the bird's turf, I sprinted through. I must have burned a hundred extra calories running full speed *away* from this bird.

Once home, I wanted to find out more about this creature and why it was attacking me. Of course, it was protecting its nest. It was a least tern, a species that travels from South America and the Caribbean to nest on Florida beaches and along the Atlantic coast. I became fascinated and concerned as I read that least tern nests

are mere scrapes in the sand and highly vulnerable. No wonder it was chasing me away. I had a lot to learn from this experience, which led me to the discovery that there were several shorebird species nesting in the area, some endangered. Some sections of the beach were even roped off to protect the birds' nests from foot traffic. I was living in a birder's paradise—I just needed a little shove to become a part of it. Would you believe it took a visit back to Brooklyn to seal the deal?

It all started with a little spiral-bound notebook. During a visit to Brooklyn in 2010, I somehow discovered it in one of the many towering book piles of an overly stuffed used bookstore. This *Birder's Life List and Diary*, published by the Cornell Lab of Ornithology back in 1986, was a journal for recording bird sightings and keeping a list—a "life list"—of all the different species one spotted. I had heard of to-do lists and bucket lists—but bird lists? I was intrigued, and the book came back with me to Florida. But it sat unopened on the coffee table for months. Little did I know that there was a lifelong passion waiting to be unleashed right there in front of me.

Though I saw the notebook on the table every day and hadn't yet used it, I never placed anything on top of it, nor did I put it on a shelf or in a drawer. This bird list was obviously something I intended to start. The daily sighting of the notebook eventually kicked me into gear, and I suggested to my boyfriend, Connor, that it was time to begin our bird list. We made a trip out to a local park to find some birds. Quite the novices, we didn't even think we might need binoculars (an essential piece of birding equipment). There we stood on a raised wooden boardwalk, ready to see birds, yet unsure how to go about it. Suddenly, one whizzed by, making a loud rattling sound. One for the list! Wait—we had no clue what species it was. Without that important piece of information, we couldn't enter it in the notebook. I noted its dull blue color and long beak, and made a point to commit its distinctive sound to memory. After several fruitless hours of straining our eyes to find more birds, we headed home.

I was intent on identifying that blue bird—there had to be a way. I thought it was a long shot, but when I typed the words "blue bird rattling sound" into Google's search box, there it was in the very first search result—a belted kingfisher. I never imagined this identification thing could be so exciting. That night, using a trusty, low-tech ballpoint pen, I carefully entered this species into the notebook; Connor and I then raised a glass of wine and toasted the belted kingfisher (this would become a ritual for all life birds we sighted together). My life list had begun, and suddenly something changed. Now, as I stared at the words *belted kingfisher* on line number 1 in the notebook, the birder switch flipped on inside me. The rest of the lines, empty, went on for several pages, all the way up to 828 (room for the roughly 700 species in North America, plus plenty of wiggle room for rare sightings). These blank lines beckoned me with the promise of hundreds of birding

View from Pier 4 beach

adventures. I had to fill as many of these in as I could. The next day, I ordered binoculars and joined the local Audubon chapter, where a couple of experts took both me and Connor under their wing. I was up, up, and away on my birding adventure.

Fast-forward to January 2012, when I moved back to Brooklyn. I was ready once again for a change of pace and some big-city opportunity. I had heard that Central Park was one of the best birding spots in the United States, but I wondered if some birding excitement was possible in my neck of the city. After two years of birding in Florida, I expected to see little more than endless flocks of pigeons and gulls in my neighborhood and hoped it wouldn't dampen my spirits.

Luckily, a friend had a room available that put me right on the Brooklyn waterfront. I was curious about the new park across the street, just up the block from my apartment. Brooklyn Bridge Park had opened in 2010. It spanned over a one-mile stretch along the Brooklyn waterfront

and extended under the Brooklyn Bridge. Sure, it was the dead of winter and freezing cold, but my pursuit of birds and the possibility of discovery had turned this former California girl into an extreme-weather trouper. Without batting an eye, I bundled up, grabbed my binoculars, and headed to the park. I spotted many gulls near the entrance, though not as many pigeons as I'd expected. I wasn't sure if that was a good or bad sign. As I approached the waterfront, the views of lower Manhattan, the Statue of Liberty, and the Brooklyn Bridge stopped me in my tracks. It sounds like a cliché, but the scene really did take my breath away. For a moment, I actually forgot I had come to see birds. This had to be the best skyline panorama in all of New York City. I knew I'd be spending a lot of time here, even if I didn't see any birds.

Of course, there would be no book if there were no birds in the park. That first winter, I encountered some captivating species. Iridescent buffleheads and spiky-headed, red-breasted mergansers were a common sight in the water between each of the park's piers. While I had observed these diving ducks many times before, I had only read about their courtship displays. Then one day I noticed two male mergansers dancing in the water. They extended their necks in unison, then, in a strikingly fluid motion, dipped their upper bodies forward and into the water. Their repeated display was obviously directed at a single female a few feet in front of them. A drama was unfolding before my eyes. Who would get the girl? Eventually she snubbed them both and swam away.

Sightings like this provided me with endless opportunities for study—and entertainment—throughout the winter. For the first time in my life, I willingly stood out in the snow for hours, watching and wondering what might happen next. I felt like a kid. I often found myself grinning ear to ear, even as I wove through a sea of straight-faced city dwellers. People probably thought I was nuts. I guess I kind of was. I was crazy for this living art exhibit.

As spring migration approached, I wasn't sure what to expect. When I lived in Florida, my spring bird lists would often number fifty to seventy-five species in a day. While I wasn't expecting those numbers in the park, I had high hopes; New York City is located along the Atlantic Flyway, one of four main migratory routes for North American birds that make their way north in spring to breed and south in fall to spend winter in the tropics. Plenty of interesting sightings were being reported in Central Park and Prospect Park, much larger parks with a wider variety of habitats. Would some of them come to my patch, Brooklyn Bridge Park? I couldn't wait to see what landed on the waterfront.

And then one day it happened. Brightly colored migrants started appearing in the young trees, shrubs, and grasses of the

Exploratory Marsh, Pier 6

park. Though not abundant and easy for any park goer to miss, they were *here*. Orioles, thrushes, warblers, and wrens were all making a pit stop in my patch. As I watched a common yellowthroat—a warbler with a beautiful black mask set against a yellow and tan head—I couldn't believe I was just a five-minute walk from downtown Brooklyn. How could this be happening in a park located right below the Brooklyn-Queens Expressway?

Something about this was even more exciting than birding in Florida. It was a challenge—sometimes I had to dig deeper to uncover these urban treasures. Each bird I found—even species I had seen daily in Florida—gave me a rush.

After a few months of birding in the park, I started to record all of my bird sightings into eBird (ebird.org), a real-time bird checklist application from Cornell Lab of Ornithology, the same organization that created the life list notebook that initially sparked my interest in birding. While I loved my notebook and continued to fill in any new species I spotted (life birds), eBird allowed me to enter a new list for every outing. It also made it easy to analyze my data and that of other birders, providing bar chart views and ways to export information for other uses. What's more, I would be playing my part in citizen science—the data reported in eBird is used by researchers, ornithologists, conservation biologists, and many others. The idea that my sightings could be used in research and conservation efforts motivated me even more to enter them in.

During spring and fall migrations, I was out most days, submitting checklist after checklist to eBird. My species count for the park was growing, and I was still grinning way too much even when I wasn't birding. When my count reached sixty, I entertained the thought of trying for 100. I was already more than halfway there—no big deal, right? Wrong. The more birds you see in a single location, the more difficult it becomes to see new species. When I first started birding in Florida, my mentor Bob told me to enjoy it, because the best

part is when everything is new. I didn't understand how right he was, as the first 100 or so birds spotted in Florida came easy, especially with Bob and his wife Lucy's help. Reaching a count of 100 in Brooklyn Bridge Park, a small park with a limited variety of habitat, was going to be tough. Still, I was determined. I committed to the quest.

I was having the time of my life on my adventure but I started to feel a bit of guilt for keeping this all to myself (apart from submitting eBird reports viewed by other birders). I wanted others to experience the magic. Birding was the ultimate urban escape, but most people had no idea that something like a gorgeous scarlet tanager was hiding in the bush they just passed. I had to let them know. I organized a group on meetup.com and started giving bird

walks in the park. Many people were just as amazed by this show as I was. It was great to see smiles appear as they observed everything from gray catbirds to ruby-crowned kinglets to yellow warblers. Their reaction made me think about how I might share this with even more people. What if I photographed birds in the park and posted them on a blog?

I had always said I would not photograph birds. I loved birding and, having several friends who are bird photographers, knew it would be a huge undertaking to get good shots of such small, quick-moving subjects. It's not that I wasn't up for another challenge; I just didn't want photography to get in the way of my birding. There's a big difference between admiring and studying a bird through a pair of binoculars and trying to get a decent photo of a bird that won't sit still. I enjoyed looking at my friends' work and followed some photographers on the Internet. I was a birder, not a photographer.

But that was then. Now there was a quest at hand, and it involved not only spotting 100 species but sharing them, too. I consulted my friend Brenda, a talented bird photographer, for advice. She said I needed a good camera, and more importantly, a big lens. I was shocked at the investment. While she warned me that this undertaking would inevitably involve some degree of frustration, she was convinced it would be one I would not regret. My mind was made up—I was going to do this—but with one stipulation. I wanted to keep my emphasis on birding. So I vowed to limit my photography to the birds of Brooklyn Bridge Park.

So off I went to B&H Photo in Manhattan, feeling guilty about my plans to make such an exorbitant purchase. I didn't know anything about taking pictures, let alone bird photography, an especially challenging type of photography. What in the world was I doing? I stood across the street from the store to collect my thoughts. Did I really want to do this? I texted Brenda for reassurance;

she gave me the OK and I headed in. B&H Photo is practically a tourist attraction, a sort of Willy Wonka's chocolate factory for photographers and videographers. A conveyer belt system, suspended high above the flurry of customers, transported top-quality gear and gadgets from stockroom to sales windows. It was overwhelming. I placed my order in the used department, wishing I could hitch a ride on the conveyor belt alongside my new toys. Eventually, I handed over my credit card at the pickup window and made it outside with a heavy bag in hand.

I took a couple of test shots in my apartment, and they looked pretty good. So the next day, I took the camera—I mean, the camera and the *giant lens*—to the park. Talk about beginner's luck. It was the peak of fall migration and flocks of red-breasted American robins and black-masked cedar waxwings feasted on berries before sunset. Several daintily spotted hermit thrushes foraged around the bases of the trees. I snapped away, and headed home to transfer the images to my laptop. I was shocked—I had captured some

Manhattan skyline from Pier 1

great shots. "This won't be so bad," I thought. Well, let me tell you: I haven't had as good a day of shooting ever since.

I got a kick out of how people assumed I was a professional photographer just because of my large lens. They would stare and make comments, with "Nice lens!" being the most frequent. Boy, was I fooling them. Only I knew that I didn't have a clue. But I took some photography classes and scoured the Internet for tutorials and tips. Eventually I became comfortable carting the behemoth lens into the park along with my binoculars. Within a few weeks, the thing might as well have been attached to my hip.

In the midst of learning about photography, my quest was going strong. Even though weeks would pass without my spotting a new species, I was inching closer to 100. I just wasn't sure I'd hit that number in this century. And I realized that, as much as I wanted photos of each and every species I saw, this wasn't a realistic goal. Bird photography requires getting close to your subjects, dealing with challenging lighting situations (such as countless leaf shadows), and freezing in action many birds that won't sit still. Plus, birds fly away. It's just what they do.

I toyed with the idea of using my own bird blind (a wall with a slit so birds can't see you). I could cart this camouflage-painted cardboard wall to the park and set it up in a promising spot. How fun it would be to hide inside with my large lens jutting out, waiting for a bird to come in for its portrait. It was a tempting prospect, but one that would surely get me arrested or banned from New York City parks. Instead, I chose to look at the situation in a positive light; the birds I couldn't photograph would still be on my list of 100 (if I got there), and they would still play an essential part in my quest.

My count edged up and up, and on September 4, 2014, I spotted a yellow-billed cuckoo, my ninety-fifth species in the park. My friend Cindy and I were facing what I call the Magical Knoll, a small slope on

Pier 1 near the East River, directly across from the skyscrapers of lower Manhattan. This is a true park hot spot, often yielding some great birds such as Wilson's warbler, northern parula, and various flycatchers. As we watched and waited, I noticed something pop its head out of the tallest tree. "It's a cuckoo!" I frantically whispered, fumbling with my camera. The bird was too far away to get a good photo, but I was able to capture a proof-of-identification shot. In less than twenty seconds, the bird was back in the tree and totally out of sight. I knew that walking over to the tree would startle the bird and cause it to take off. It had stopped in the park to rest and fuel up for its continued journey south, and that was more important than any photo. I never saw that cuckoo again.

Ten days later, I was up to ninety-nine species. It was a high point in migration season, and over the past week and a half I had a lucky streak (helped no doubt by my spending about five hours in the park each day). I spotted a great-crested flycatcher in the same tree I had seen the cuckoo, as well as a veery, a Cape May warbler, and a peregrine falcon. What would my 100th bird be? Only now could I seriously contemplate this question. I wanted it to be something special. For a time, I wished that coveted cuckoo had been number 100. There were still some common New York City birds that I hadn't spotted in the park. Black-capped chickadees, for example, were abundant in Central Park. They preferred tall trees, and most of the trees in my patch were still too young to attract them. Still, it was possible that one would show up and that the 100th bird would be a pretty common one. Why was I torturing myself? Wouldn't a chickadee be good enough? I should have been happy to even reach ninety-nine species, and ecstatic if I got to 100, no matter what the species.

As it turns out, I didn't have to settle.

Brant geese perched on pilings

On September 16, 2014, I noticed a bird in the trees of Pier 1. At first it looked like a warbling vireo, which was already on my park list, but as I got closer, I believed what I was seeing was a Philadelphia vireo. I say "believed" because, in birding, when you want a species badly, it's easy to see things that aren't really there. A Philadelphia vireo would be a rare sighting, not only for the park, but also for New York City. Even though the bird was in the shadows and moving constantly, I managed to get a few proof-of-identification pictures. When I got home, I pored over the photos. Their low light and shadows made it hard to tell what I was actually looking at. I started to doubt myself. Had I been talking myself into this species? Believe it or not, I wouldn't be the first one to have ever done this. Expert birder Kenn Kaufman addresses this phenomenon in his book *Field Guide to Advanced Birding*: "Our subconscious drive to find a rarity can genuinely alter our perceptions, so that our minds . . . ignore things that don't fit in with what we want the bird to be."

Kenn was right. I was obviously not in my

right mind and needed backup. I sent the photos to my expert bird identification consultants Bob and Lucy in Florida. Within minutes, Lucy replied, saying there was no doubt it was a Philadelphia vireo because of the bird's eye line (a thin dark line of feathers on both sides of or behind the bird's eye). I had reached my goal! And my 100th bird was a rare sighting. It was almost too good to be true. Looking back, I should have had cake to celebrate.

My quest was now complete—or was it? I kept going, of course, and the black-capped chickadee showed up as my 108th species just a few weeks later. My count is currently at 134 and I'll continue the quest as long as I'm able to get out to Brooklyn Bridge Park. At some point, I know years will pass before I spot another species there, but that doesn't bother me at all. Each day I visit the park, I learn more about birds. And I witness common birds doing uncommon things, which is often even more exciting than spotting new species. (My favorite sighting to date was a northern mocking-bird—a relatively common park bird—that stopped singing, regurgitated a berry, digested it, and then continued to sing.) My quest is not just to discover new species along the waterfront; it's also to learn as much as I can about birds, continue to foster my appreciation of them, and to share them through my photos. Though when that next new species finally perches in my patch, I admit I'll be grinning for weeks.

The bird species accounts and photos in this book are organized by the season in which the photo was taken and/or in which the story of the sighting took place. Within each season, the birds appear roughly in the order they were first sighted during my quest. (If you're interested, you can see exactly how my adventure unfolded on page 248, where I've listed every species in the order I first saw them in the park.) Many birds can be sighted in the park year-round or in multiple seasons, though some migrants show completely different or drabber plumage in fall than in spring (e.g., the male chestnut-sided and

Dark Forest, Pier 1

blackpoll warblers). For information on the park's year-round resident birds and when to expect certain migrants, you can take a look at Birds Commonly Sighted in Brooklyn Bridge Park on page 267.

I created this book not only to share my journey, but also to inspire you to take a closer look outside your home or office window, even if it looks out over an endless metropolis. Whether you live in Brooklyn, have visited, or have never been, I hope this book will delight you and lead you to discover beautiful birds and other wildlife that exist in your city.

Ring-billed gulls, Pier 6

WINTER

The park is blanketed in snow. Mounds of ice drift between the piers. Farther out in the distance, a loon swims in the East River. Two of the park's common winter diving ducks—bufflehead and red-breasted merganser—swim closer to shore, navigating around the frozen maze. At the Pier 3 uplands, a song sparrow extracts seeds from tufts of little bluestem grass that peek out from the fluff. A white-throated sparrow forages along the wooded paths of Pier 1. I watch as it licks ice from its bill after having dug below the snow.

While the pace of winter birding is slower than in spring or fall, there are fewer places for birds to hide. Sparrows can be seen foraging under leafless shrubs. A northern mockingbird is easily located by sight rather than sound; its silhouette stands out on a tree barren of cover. The many ducks of winter are easily spotted as they dabble and dive in the water between the piers. Instead of scanning the trees for migrating warblers, I now turn my attention to the water and the possibility of rare waterbirds—loons, grebes, or ducks.

Gadwall, Pier 4 beach

GADWALL

(Anas strepera)

The gadwall is a waterfowl of park winters, when it joins the other dabbling ducks in their daily parade through the pilings of Pier 1. From a distance, it appears rather drab, with a smaller and darker bill that barely distinguishes it from the mallards and American black ducks that paddle beside it. But when the male gadwall comes ashore at Pier 4 and begins to preen, there is no mistaking its peerless plumage. The sight brings me to a halt; I study and admire and don't want to blink.

In summer, some gadwall breed in freshwater and brackish marsh habitat in the city and throughout the Northeast. Others migrate farther north or to the Great Plains and Canadian prairies. While they currently don't nest in Brooklyn Bridge Park, that could change; a gadwall pair had me hoping as they moved into one of the park's freshwater ponds and lingered well into summer before finally departing in July.

Gadwall, Pier 1 pilings

ROCK PIGEON

(Columba livia)

The rock pigeon is a member of the dove family, a group that has connotations of sweetness and peace. Yet rock pigeons have a bad reputation—quite different from that of their cousin the mourning dove. What seems to bother people most about pigeons is their sheer numbers, and they're notorious for pecking at our feet and pooping on our stoops. But in Brooklyn Bridge Park there aren't as many pigeons as in other city parks—I typically see only a dozen or so on my walks along the waterfront. Partly this is because people don't seem to feed the pigeons in this park. It also has to do with the bird's preferred roost, which in its native Eurasia is high atop sea cliffs and ledges. In the city, rooftops and decorative ledges provide the perfect substitute. Such ledges are nearly absent in Brooklyn Bridge Park, resulting in a nicely balanced pigeon population.

For as many people who disdain the city pigeon, there seem to be just as many that adore it. I was delighted when this photo elicited many *oohs* and *aahs* and became one of my most popular. I found this pigeon—I mean, lovely dove—resting on a bike rack by the Pier 6 volleyball courts.

Rock pigeon, Pier 6

White-throated sparrow eating sumac, Pier 1

WHITE-THROATED SPARROW

(Zonotrichia albicollis)

I n winter, when every leaf has fallen from every tree, the fuzzy red fruit clusters of the staghorn sumac remain, providing food for the park's resident winter birds, including the white-throated sparrow. But this sparrow—named for its white bib—is more often seen (and heard) foraging on the ground, doing its characteristic double-hop through dead leaves in search of hidden insects.

The white-throated sparrow's loud and strikingly clear melodic whistle—*oh, sweet Canada, Canada, Canada*— lasts a full four seconds. On cold winter mornings when no one else is in sight, I find the song takes on an eerie quality that's in tune with the still and frozen landscape. When spring finally arrives, the exact same song seems much cheerier.

White-throated sparrow, Pier 1 wooded paths

BRANT

(Branta bernicla)

I n winter on the lawns of Pier 1, the sunbathers of summer are replaced by a foraging flock of brant geese, plucking strands of dead grass amid patches of snow and ice. Come sunset, the honking begins as the geese take flight, circling clockwise from Bridge View Lawn down to the pier pilings, up the East River, and back again. For people in the park in the dead of winter, the flight of the brant offers quite a show. The flock, more than two dozen, circles over Pier 1 and the East River, passing in front of some of the city's most recognizable landmark scenes: the Statue of Liberty, the lower Manhattan skyline, and the Brooklyn Bridge.

In summer, brant migrate farther north than any other goose, traveling in large flocks to the high Arctic tundra. Though some nonbreeders remain in the city through summer, only in winter can you catch their Pier 1 flight spectacle.

Brant and Statue of Liberty, view from Pier 2

Red-breasted mergansers, Pier 4

RED-BREASTED MERGANSER

(Mergus serrator)

cicles hang from the baby blue metal umbrellas that shade the picnic tables south of Pier 4. Piles of snow blanket the benches beside the barbecues. And yet I don't wish for warm weather, for a summer past or a summer to come. Now, it is winter—the season for diving ducks. In the water just offshore, I watch a red-breasted merganser descend to its underwater hunt. A moment later it surfaces with a small crab in its bill. Unlike the many migrants of spring and fall that forage in the tops of trees, winter ducks are easy to observe in the water below the piers (and much easier on the neck).

In its breeding plumage, the male red-breasted merganser is the most ornate of the park's common winter ducks, with its sleek body, long bill, and iridescent head feathers. While females sport the same long, serrated, red-orange bill, they lack the contrasting plumage and have a rust-red neck, slate-gray back and breast, and white belly.

BUFFLEHEAD

(Bucephala albeola)

As winter approaches, I eagerly await the arrival of the bufflehead, a small duck that breeds in boreal forests of Canada and Alaska and winters in the United States. Weeks before it can reasonably be expected to arrive, I can't help but scan between the piers for my favorite diving duck. By the time the first one makes its appearance in November, it feels like a lifetime since I've seen a bufflehead. Just as I bring the bird into focus, it propels itself up and forward for an elegant dive, one that doesn't seem possible for such a chunky duck. Down it goes, its tail feathers an outstretched fan entering the water. While I wait for the bird to surface, I imagine its hunt for a mollusk or crustacean below. What's it really like down there in the depths of the East River? Maybe I don't want to know. After twenty seconds or so, the bufflehead bobs to the surface like a rubber duck.

More buffleheads arrive in the park late in fall, maxing out at about a dozen. Once they settle in, the real show begins as males engage in courtship displays. The sound of a stuttered splash directs my attention to the Pier 1 chase; amid the pilings, a male is on guard and impresses his mate by running off an encroaching male. The successful defender races around the water and bobs his head forward and back. Comical.

After a near six-month run, the bufflehead show closes in spring. These engaging ducks depart the park for their northern breeding grounds, where they will nest in abandoned northern flicker holes.

Bufflehead, Pier 5

Bufflehead diving, Pier 5

Bufflehead (female), Pier 4

Downy woodpecker, Pier 1

DOWNY WOODPECKER

(Picoides pubescens)

Something about the way a downy woodpecker lands on a tree brings to mind a parachutist or perhaps a flying squirrel. An imaginary *swoosh, swoosh, swoosh* sounds in my head as I watch it fly toward the next tree trunk in what seems like slow motion; its wings seem to beat much too slowly for it to stay airborne. The landing is smooth. Without the least bit of hesitation or need to gain footing, the bird grasps the vertical surface with its zygodactyl feet (two toes in front and two in back, unlike the three in front and one in back of most birds) and props its tail against the bark for support. Just by the simple act of landing—simple for a bird, at least—the downy woodpecker exhibits grace, efficiency of movement, and the utmost coordination.

One of my favorite photos, the shot on the opposite page was taken after a heavy snowfall that turned the park into a winter wonderland. The tiny snowflakes on the bird's bill were likely picked up as it extracted insects from snow-covered branches.

Downy woodpecker, Pier 1

RED-NECKED GREBE

(Podiceps grisegena)

There are many ways birders communicate rare or interesting bird sightings—listservs, eBird, even rare bird alert phone hotlines. But Twitter? It turns out it really is (sometimes) for the birds. I found out about this red-necked grebe when fellow birder and Brooklynite Michael Yuan tweeted his park sighting. Though technically not rare for this time of year, the red-necked grebe shows up only a handful of times during winter, so New York City birders jump at the chance to see one. And for me, this was not just a park bird; it was a lifer (aka life bird). During the following six weeks, I enjoyed watching the grebe grapple with marine worms as it continued to make appearances between the piers.

The red-necked grebe breeds in Canada and Alaska during summer when it sports its ornate breeding plumage—reddish brown on the neck, white or gray on its cheeks, and a black cap. During courtship, males and females often exchange gifts of water weeds and dance together by racing upright across the water in unison. Their fragile nests, which float atop patches of aquatic plants, are fully surrounded by water.

Red-necked grebe, Pier 5

Red-throated loon, Pier 5

RED-THROATED LOON

(Gavia stellata)

On a quiet and cold winter morning, I hear the tranquil splash of a perfect dive executed just north of Pier 5. As I wait for something to surface, the still and reflective water offers no hint of the drama occurring below. And there's the sound again, but this time in reverse, as the red-throated loon breaks the surface, with water droplets beading on the feathers of its head and neck. Within seconds, the water is still again, with the loon floating gracefully atop.

The red-throated loon usually shows up during the coldest months, when it can be seen in the East River, and sometimes between the piers and along the park edges, diving for fish, mollusks, and crustaceans. Its striking breeding plumage—which includes a red throat patch, dark gray head and neck, and vertically black-and-white striped nape—is usually visible only in the Arctic, during breeding season. In its drab winter plumage, the red-throated loon's slightly upturned bill helps to distinguish it from the common loon, another winter visitor, which tends to hold its bill in a horizontal position.

WINTER WREN

(Troglodytes hiemalis)

Walk into the Exploratory Marsh on Pier 6 and you'll be transported from the fast pace of New York City to a tiny oasis. Designed for kids to discover birds and butterflies, this 5,000-square-foot circular garden has turned up some amazing finds, including the black-throated blue warbler, wood thrush, and red-eyed vireo. And in November 2014, this winter wren stopped by for a two-week stay.

Just because the winter wren has *winter* in its name doesn't mean it's a sure sighting of the season. It's known for being highly secretive and difficult to spot, even when you can hear its vibrant song emanating from somewhere right in front of you—usually in a cluster of dead wood or brush. Here's a scene that replayed daily that winter: I enter the marsh and see a tiny flash of rusty brown disappearing into a dense thicket just a few feet away. I watch. I wait—every day for a week—until I give up all hope of photographing the wren. (I admit I harbor some ill will toward the little thing.)

A few days later, on one of our birding-and-brunch Sundays, my friend Bob and I were on our way to the marsh when we noticed movement in the tall conifer at the center of the adjacent water playground. Moving quickly between the levels of the sparse tree was not one, but two winter wrens—and with few places to hide. This one hopped down to the metal railing at the base of the tree to finally give me a good (and quite adorable) look.

Winter wren, Pier 6 Exploratory Marsh

Canvasback, Pier 2

CANVASBACK

(Aythya valisineria)

As I wove through the park's snow-shoveled paths on a morning run, surrounded on either side by a foot of fluff, I tried hard to resist the temptation to scan the water for winter ducks. Then Pier 4 came into view and reminded me of one that got away—a male canvasback. This large and striking red-eyed duck had recently been spotted on the beach but was gone by the time I arrived. Thanks to eBird, I knew these ducks were being sighted just across the river at Liberty Island. What would it take for one to grace me with its presence?

Giving in to my birding urge, I took the low path along the water to watch while I ran. Moments later, I was stunned. There, between piers 2 and 3, were three canvasbacks. They were sleeping, floating peacefully with necks curled in, bills buried under feathers, and one eye open. I ran down to the railing and admired the birds through nothing more than a pair of eyeglasses (and vowed never to leave my binoculars at home again). One bird's red iris peeked out from its feathers. It appeared to catch an extra glow from the warmth of the winter sun. I was torn for just a moment, then quickly raced home to grab my gear. But there was no need to hurry; I returned to find all three still asleep. Soon I had hundreds of photos of canvasbacks—sleeping canvasbacks. Would one ever wake up and raise its neck? Three hours later, they finally started to stretch, rising up from the water and flapping their wings. After the action, I watched as they paddled out, lifted off, and headed south out of sight.

NORTHERN SHOVELER

(Anas clypeata)

As I rounded the end of Pier 5, I saw a small patch of the brightest white in the water near Pier 4's Bird Island. I raised my binoculars to focus on my 111th park species, this male northern shoveler. The male's gleaming white throat and chestnut flanks make it easy to identify, even from a distance. But up close, the duck's wide, spatulate bill is the main attraction. (Female shovelers have these bills as well but have drab grayish-brown plumage.)

These ducks behave a bit differently than most other members of the genus *Anas*. Most *Anas* males abandon the female immediately after she lays her eggs, but the seemingly kinder (actually more territorial) shovelers wait until sometime later during the incubation period. And while most dabbling ducks forage by upending headfirst in shallow water, shovelers do this only occasionally; more often they propel themselves forward while scooping just under the water's surface with their large, specialized bills, filtering food from water with tiny projections—called lamellae—that line the bill.

Northern shoveler, Pier 4

Ruby-crowned kinglet, Pier 1

SPRING

The chill of winter hangs on as spring approaches. The sparrows are the first to stop by on the way to their northern breeding grounds. A fox sparrow and dark-eyed junco are the earliest arrivals.

The cold finally breaks and tiny buds appear on the shrubs and trees; a ruby-crowned kinglet checks them for newly hatched insects. Early-season warblers and vireos are spotted easily as they forage in the still leafless trees. But most birds stop only briefly for food, then continue on to their breeding grounds where their tasks are many: attract a mate, stake out a nest site, build a nest, raise young. Through all of this, they must defend themselves—and their mates and young—from predators.

Spring migration peaks in May, and the flurry of birds passing through the park ends by June. Yet two spring arrivals stay—the barn swallow and the gray catbird.

MALLARD

(Anas platyrhynchos)

Although this mallard looks like it's ready for a ballet, it's just doing a "comfort movement" enjoyed by many birds, the wing-and-leg stretch. (Other examples of comfort movements include scratching, shaking, preening, and bathing.) When I came across this duck on Pier 5, it was perfect timing, as I had been studying Kenn Kaufman's *Field Guide to Advanced Birding*, learning about avian plumage variations, molt, and feather arrangement. I jumped at this opportunity to count the mallard's primary wing feathers—a set of long, stiff feathers used for flight. Most bird species, including the mallard, have ten of these primaries, numbered from P1 to P10, with P10 being the outermost wing feather. (You can count these in the photo, though P1, the innermost primary, is partially hidden beneath one of the bird's secondary feathers. Secondaries are another set of feathers that are also used in flight.)

A few months after I spotted this mallard, I witnessed another feather event as four male mallards transitioned from breeding plumage—with that striking, iridescent green head—to drab brown eclipse plumage similar to that of the female. This molt included replacement of flight feathers, leaving the ducks unable to fly for several weeks. I watched as they stuck together for safety, often swimming near the Pier 4 beach or resting on the partially submerged railroad float transfer bridge extending out from Bird Island.

Mallard, Pier 5

DOUBLE-CRESTED CORMORANT

(Phalacrocorax auritus)

While it's exciting to spot new lifers or birds that are unfamiliar, it's also comforting to have old friends around. So when I moved back to Brooklyn, it was nice to find a species I had come to know well in Florida—the double-crested cormorant. There it was, ready to greet me, perched on the pilings at Pier 1. Cormorants are year-round city residents that use Brooklyn Bridge Park as a daytime roost and fishing spot. The park's exposed pilings are the perfect place for these large waterbirds to rest and dry their wings in between their deep dives (of up to twenty-four feet). Cormorants swim with most of their body submerged; their necks can be seen jutting up from the water as they fish around Pier 1. Just before sunset, the birds depart for their nighttime roosting locations around the city.

One spring I arrived at Pier 4 beach to find two cormorants engaged in courtship on another of their favorite perches, a metal arch sticking up from the water (since removed, sadly). The birds extended their necks forward and straight, bills open. One dove underwater and surfaced a minute later with a reed in its bill. It sidestepped carefully up the arch with its large, webbed feet. When it reached the top—where its potential mate awaited—the cormorant presented the dripping reed. I watched as they playfully passed it back and forth. Alas, the reed fell back into the water and the male jumped down to try again. In this courtship display, the male was presenting the female with nesting material.

Since cormorants don't nest in the park, I'll never know how this spring romance progressed. (Cormorants nest on some of the city's small, isolated islands, including two in the East River—South Brother Island to the south and Belmont Island to the north.)

Double-crested cormorant, Pier 2 kayak launch

Eastern phoebe, Pier 1 Vale Lawn

EASTERN PHOEBE
(Sayornis phoebe)

The lawns of Pier 1 are surrounded by aesthetically pleasing fences made of smoothed, slender tree trunks, the poles connected by strands of thin wire. During migration, this low-lying wire is the preferred perch of the eastern phoebe, where it can easily eye its prey—flying insects buzzing inches above the grass. I watch as it waits for the perfect fly while pumping its tail in rhythm (tail-pumping is characteristic of this species). In a sudden burst of flight, it sallies out over the lawn and snatches its snack, unaware that it is likely saving some city dweller from another itchy bug bite.

AMERICAN ROBIN

(Turdus migratorius)

A year-round park resident, the American robin provides endless opportunities for observation. In spring, robins blanket the lawns of Pier 1, staking out their nesting locations and defending their territory, until every tree on Bridge View Lawn has a robin or two patrolling below it. If another encroaches—setting just one tiny toe over the invisible territory line—it is immediately chased away and scolded with a loud series of *yeep* notes. Later in spring and throughout summer, robins can be seen delivering food to their young in the nest and feeding fledglings on the grass.

American robin on the steps of Harbor View Lawn

American robin, Pier 4

House sparrow after snatching ephemera, Pier 1

HOUSE SPARROW

(Passer domesticus)

f you're not a birder and aren't sure what that small brown bird on the ground is—the one eating the crumbs of your bagel or, in my case, chocolate chip cookie—house sparrow would be a good guess (at least in urban and suburban settings). When this nonnative species was brought from England to New York City in 1852, its population exploded so fast across the United States that it quickly entered pest status. The house sparrow is not protected by the Migratory Bird Treaty Act and is one of the few birds that can be legally exterminated.

The great thing about consistently birding the same patch is that you're there in the good times and the bad. In birder's terms, the good times are those days during migration when there are too many birds to identify. And the bad—not a bird in sight. But in my urban patch, I could always count on seeing a house sparrow. Inspired by Jon Young's *What the Robin Knows*, a book in which he encourages closer study of everyday birds, namely the American robin (no birder would recommend studying pests), I decided to make the most of slow birding days by studying house sparrows.

It turns out that house sparrows—many a birder's most-hated species—have picked up the behaviors of birds every birder admires. I've seen them acting like warblers—flitting quickly along the branches high in a tree like a yellow warbler, acting secretive like an ovenbird, wading like a waterthrush (sans tail-bobbing, though I wouldn't put it past one to do so). They've impressed me by scaling trees using their tails for support, a behavior characteristic of woodpeckers and creepers. But most impressive was the weeklong fly-catching extravaganza on Bridge View Lawn one spring.

This event was not headlined by actual flycatchers—eastern phoebes, *Empidonax* species, or great-crested flycatchers—but by house sparrows. Hundreds, if not thousands, of tiny insects—likely of the mayfly genus *Ephemera*—swarmed above the lawn. Taking off from the grass and low branches, the house sparrows hovered about three feet above ground, snatching the newly hatched insects in midair until their bills were chock full o' flies. But it didn't appear that the birds were eating them or delivering them to young. One female had so many flies that a few fell from her bill onto the grass. She then proceeded to stuff them back into her mouth—still with no sign of actually eating any. Maybe the birds were taking this opportunity to perfect their fly-catching skills. It certainly provided the ultimate training ground but lasted only a couple of days, because the flies that weren't snatched soon died a natural death—these ephemera live only a day or two after hatching.

NORTHERN MOCKINGBIRD

(Mimus polyglottos)

When I heard the mimicry of a mockingbird at the park's Pier 6 entrance, I decided to play what proved to be quite an entertaining game. I listened to the bird's vocal repertoire, trying to identify as many different birdsongs as I could. The easiest to pick out was the unmistakable namesake call of the killdeer, a sound I got to know very well when I lived in Florida. As the mockingbird continued to sing, I heard it imitate the sweet song of a northern cardinal, the wail of a blue jay, and even the cute *witchety-witchety-witchety* tune of the common yellowthroat, the park's most common migrating warbler. Playing *Name That Tune* with the mockingbird reminded me that I needed to work on my birding by ear; there were at least ten calls I couldn't identify.

Northern mockingbird,
Pier 3 uplands

Northern mockingbird,
Pier 3 uplands

EUROPEAN STARLING

(Sturnus vulgaris)

After a long day of birding, I was sitting on an old mooring post at the park's Pier 1 entrance, enjoying one of my favorite indulgences—a hot fudge sundae with coffee ice cream from Brooklyn Ice Cream Factory—when this starling went cuckoo on me, poking its head out of a vent hole in a snack shop. It flew off and was back in the hole a few minutes later, making a food delivery to its young hidden inside. Over the next few weeks, I visited often and eventually got to see what I was hoping for—a small nestling poking its head out from the hole.

The European starling first touched down in the city back in 1890, but it didn't fly here. Along with fifty-nine others, it arrived to Central Park in a cage carried by Eugene Schieffelin, a man on a mission to share the birds mentioned in Shakespeare's works with the New World. Schieffelin imported the starlings from England. At the time, the ecological impact of such a pursuit was not known. The starling population grew rapidly to pest status and now has an estimated population of over 200 million.

European starling,
Pier 1 entrance

*European starling nestling,
Pier 1 entrance*

European starling, Pier 6

Northern cardinal, Pier 1

NORTHERN CARDINAL

(Cardinalis cardinalis)

The northern cardinal is one of the city's most common birds. Its loud, distinctive chip note can be heard on a stroll through any Brooklyn neighborhood. Whenever I visited nearby Cobble Hill Park—a small park I pass on the way to the subway—a cardinal would perch at eye level just a few feet in front of me. But in Brooklyn Bridge Park this species eluded my lens for nearly a year. And I'm amused—and frustrated—that I have captured much better shots of the marsh wren and the ovenbird, birds that are notoriously secretive and difficult to photograph. I dream of the day when I find the park cardinal in perfect light, perched on a branch blanketed with snow. But even on a gray, overcast day, on a bare branch, this bird shines.

CANADA GOOSE

(Branta canadensis)

t was the first of April, and a flock of Canada geese calmly plucked grass from Harbor View Lawn. I stood on the adjacent path, snapping some standard and somewhat blah goose photos. Looking through the lens and focusing on a single bird, I was now unaware of anything but me and the large waterfowl. Without warning, the goose became so agitated (though I had no idea why) that it stuck its tongue out in an appropriate April Fool's expression. As I lowered my camera, I saw the culprit—a young woman snapping smartphone shots of the bird from just a few feet away. (Geese and swans are known to attack people when they get too close.) The woman saw my concerned look and backed off, at which point the goose went back to feeding. I guess you could say someone else did the dirty work to capture this funny photo.

Canada goose, Pier 1 Bridge View Lawn

Mourning doves, Pier 1

MOURNING DOVE
(Zenaida macroura)

Extending out from the small beach at Pier 4 is the park's small nature preserve called Bird Island. Before the preserve was completed in the summer of 2014, I would see perhaps two mourning doves throughout the entire park. That following spring, I noticed two mating on the south end of the island. Though I never saw a nest there, there were likely several. In late spring, I spotted six doves. About a month later I regularly saw eleven. And in August I counted a whopping twenty doves foraging on the island's north end. Bird Island had become Dove Island. This mini population explosion is not common among native birds, but mourning doves often raise an unusually high number of broods—up to six each season.

Mourning dove, Pier 1 Bridge View Lawn

PALM WARBLER

(Setophaga palmarum)

A flash of yellow pops up on Harbor View Lawn. It's the palm warbler, now bobbing its tail as it moves through the grass, checking each blade for insects. It hops above the blades to grab something—maybe an aphid, ant, or fly. I try to imagine the bird's view; the grass must appear as an overgrown jungle. Unknown danger might lurk around any corner. When the warbler hops above the maze, does it notice other birds around? Does it see me? How does it not get dizzy constantly checking up, down, and around for food? And then I realize . . . how hard it is to be a bird.

Often the palm warbler perches on the thin wire fences surrounding the park's lawns, where it can be spotted from a distance, constantly bobbing its tail. In spring, it stops in the park on its way to Canada or northern states in the United States to breed. In winter it's destined for points closer than those of most warblers—the southeastern United States or the Caribbean.

Sometimes birds have names that don't seem to fit. The palm warbler, a bird of many open habitats, was named in 1789 when it was found foraging low among the palm trees on the island of Hispaniola.

Palm warblers, Pier 1 Harbor View Lawn

Palm warbler, Pier 1 Vale Lawn

Dark-eyed junco, Pier 1

DARK-EYED JUNCO

(Junco hyemalis)

hear the sound of trilling whistles as a flock of juncos passes overhead. The birds' white outer tail feathers flash prominently in flight. As they land on the lawn, I start to count; over fifty are foraging in front of me on Bridge View Lawn. Other fall sparrows—white-throated, chipping, and song—are now greatly outnumbered.

Not only is the dark-eyed junco one of the easiest sparrows to identify, it's also one of the most common birds in North America, where its population is second only to the American robin. Juncos have a dark gray body—females are browner above—and a plain white belly. In the western United States, there are different subspecies of dark-eyed juncos, each with different color variations.

In winter, junco flocks have a dominance hierarchy—a rather sexist pecking order—that determines who gets to eat first. (Can you imagine?) Adult males are at the top, then juvenile males, followed by adult females and, finally, juvenile females.

FOX SPARROW

(Passerella iliaca)

n winter, the fox sparrow stands out against the uniformly drab dead leaves beneath the park's thickets. Its foxy brown upper parts are characteristic of the "red" eastern subspecies (*Passerella iliaca iliaca*) with which Brooklyn is blessed. (The three western subspecies look similar but with dingier plumage colors that range from brown to gray.)

The fox sparrow forages using the "double scratch" movement—using both feet simultaneously, it takes a single hop backward through the leaves to reveal insects and seeds below. (When I have too much caffeine, I often imitate this move while listening to jazz.) Other park birds that do the double scratch include the white-throated sparrow, dark-eyed junco, and eastern towhee. One of the best ways to find the fox sparrow is to listen for scratching sounds in the leaf litter during winter. Most of the time, the search reveals the more common white-throated sparrow, which I'm still delighted to find; spotting a fox sparrow is icing on the cake.

Fox sparrow, Pier 6

*Common grackle and
nestlings, Pier 2 sport courts*

COMMON GRACKLE

(Quiscalus quiscula)

High above the covered basketball courts on Pier 2, this common grackle nest likely went unnoticed for weeks. While walking along the main park path, I spotted an adult grackle flying in and out of the sports facility. Its flight seemed so urgent—could it be delivering food to its young? As I headed onto an empty basketball court to investigate, the bird flew past and up to the highest point of the awning. I noticed some twigs hanging over the beams; it had to be a nest. I backed up slowly in anticipation until my vantage point revealed this nest full of hatchlings, begging to be fed.

During spring and early summer, common grackles frequent the Pier 1 paths where they mate and forage. They can occasionally be seen engaged in their "ruff-out squeak" display that starts with a puffing out of feathers (almost like an inflating balloon) and culminates in a tinny, electronic-sounding yelp.

Common grackle, Pier 1

EASTERN TOWHEE

(Pipilo erythrophthalmus)

Never underestimate the power of the New York City pedestrian. After getting occasional glimpses of this eastern towhee in the shrub stand at the top of Vale Lawn, I watched it drop down into the thicket and out of sight. I could hear it hopping and scratching in the leaves beside me, under a tangled shrub so low to the ground that the bird was impossible to see. A trio of park visitors came into view, walking up the path straight toward the totally obscured towhee and the visible me. I backed up in anticipation of what I knew would happen next—as the people approached the towhee's foraging spot, they startled the bird, causing it to fly up to a branch in plain sight and perfect light. Lucky breaks like this also happen with people walking their dogs. Even if it doesn't present a photo op, it can tip me off to a bird's presence in the area when I see something fly away.

The eastern towhee is a large and boldly colored member of the sparrow family (Emberizidae). The female of this species has the same rufous and white markings as the male, but its upper body, head, and tail are a warm brown rather than black.

Eastern towhee, Pier 1

Great black-backed gull, Pier 1 pilings

GREAT BLACK-BACKED GULL

(Larus marinus)

While it's always exciting to add a new species to my park list, it's even better when it's a life bird. Even so, I never expected to get giddy over a gull, but that's what happened when I spotted my first ever great black-backed on Pier 1. On one of the exposed pilings, the bird's bright white plumage, draped in the darkest of blacks, stood out like a king's cloak, outshining the many pale-gray ring-billed gulls that surrounded it. More than two years have passed since that gull became a life bird, but even daily sightings of it haven't dampened its beauty. I have yet to pass one by without taking an admiring look through my binoculars.

The great black-backed gull is the largest gull in the world with a wingspan of over five and a half feet. This one is performing what's called the "long call," an aggressive sound signal accompanied by a raising and lowering of the head.

BROWN THRASHER

(Toxostoma rufum)

A stream of reddish brown flies low over Bridge View Lawn. It's a brown thrasher—a striking and camera-shy species that makes a brief stop on Pier 1 during spring and fall. Even before the trees and shrubs leaf out, the bird still manages to stay hidden. On one occasion, it perched deep in the tangles of a shrub with only its eye visible. I admired the bird's bright yellow iris and waited for it to move. But it didn't, at least not during the twenty minutes I had my binoculars focused on its eye.

When the thrasher appeared that same fall, I knew I had to come up with a plan. I sat along Vale Lawn behind a small fence post, hoping to appear to be a part of it. This worked like a charm. The thrasher hopped out from under the trees lining the Magical Knoll and took a look around. I held my breath as it hopped up the lawn and foraged right in front of me. As I left the area, I saw that the thrasher had a friend—the two long-tailed beauties flew into a sumac and began to eat the fruit.

The brown thrasher is a member of the family Mimidae—which also includes the gray catbird and northern mockingbird. It is known for having a large repertoire of over 1,000 song types. While mockingbirds make three or more attempts at a musical phrase, the brown thrasher tends to repeat the same phrase only twice.

Brown thrasher,
Pier 1 Vale Lawn

Herring gull, Pier 4

HERRING GULL

(Larus argentatus)

t's not often I stop and watch a herring gull in the park—they are usually flying by or resting on a distant piling. But this one grabbed my attention at Pier 4 as it kept bracing itself against the surf while standing on a low post. Time after time, it would suffer a splash of East River water and fall backward. But it didn't seem to mind; it always hopped right back up to await the next incoming wave.

Herring gulls can be spotted in the park and along the city's shorelines year-round. Their large size helps distinguish them from the park's smaller and more plentiful ring-billed gulls. Adult herring gulls have a red patch on their lower mandible (bill)—called the gonydeal spot—that serves as a sort of "food button." When a hungry chick pecks on the spot, it stimulates the parent to regurgitate food to feed it.

BLUE JAY

(Cyanocitta cristata)

Outside the park, I encounter a blue jay daily. I hear its shrill call on my way to the subway. I catch a glimpse of its blue, white, and black plumage as I lug bags of groceries back to my apartment. But I rarely come across a blue jay in Brooklyn Bridge Park.

How could this familiar backyard bird be nearly absent? It's possible that the blue jay prefers the nuts of more mature trees. One of the jay's favorite snacks is an acorn, a nut produced only by oak trees at least twenty years old, an age that some of the new park's oaks have yet to reach. But the surrounding neighborhoods, lined with hundred-year-old oaks, are overflowing with the jay's favorite Brooklyn bounty. Acorns, along with other nuts and seeds, make up over 75 percent of the blue jay's diet.

Lucky for me—but unlucky for the park's robins—a blue jay occasionally flies down from Brooklyn Heights in search of higher-protein fare in the form of a robin's egg or (gasp) a robin nestling. The jay you see here was eyeing a nearby robin's nest on Pier 1's Granite Prospect. I'm happy to report that, right after I took this photo, the adult robins chased the still-hungry jay out of the park. In fact, over time I would see that the parents' diligent defense resulted in the entire robin brood fledging successfully.

*Blue jay, top of
Granite Prospect*

Barn swallow, Pier 1 Long Pond

BARN SWALLOW

(Hirundo rustica)

Once I decided to write this book, I made a list of the photos I hoped to capture in the park—*really* nice-to-haves. At the top of the list was a photo of a barn swallow gathering mud for its nest, a behavior I had witnessed at the Long Pond of Pier 1 in my pre-camera days. This highly anticipated photo op even earned a spot on my Google calendar, where I entered the estimated date of barn swallow arrival. Weeks before the date I watched the pond intently, worrying way too much about water levels and availability of park mud. I breathed a sigh of relief around mid-May when I came upon the Long Pond and saw that the water levels had receded. I took a seat on the bridge overlooking the pond—and the mud, that heavenly mud—and waited patiently. Within a few moments, I heard the bubbly squeaks of the barn swallows; they landed and began to gather the good stuff.

Barn swallows build nests under the park's piers. They combine the mud with dry grass to create a cup-shaped nest and line it with feathers. By July, young fledglings are perched on the railings and ridges of the piers, awaiting a feeding from their parents. Adult birds can occasionally be seen mobbing and dive-bombing any crow that dares to hunt for a barn swallow brunch. (Crows do eat young barn swallows, both in the nest and out.)

Barn swallow (juvenile), Pier 2

BLUE-HEADED VIREO

(Vireo solitarius)

As I was leaving Pier 1, I saw a small flash of white in a shrub next to the fresh-water marsh. Ruby-crowned kinglets and common yellowthroats had been frequenting this mini-patch, but neither species sported such a gleaming shade of white. The color was now hidden, but I saw movement; I edged toward it, hoping for a surprise. The bird then foraged its way into plain sight. It was a blue-headed vireo, one of my favorite species, looking at me through its characteristic spectacles—white eye-rings joined by a white band just above the upper mandible.

Vireos look similar to warblers, but most have chunkier heads and thicker bills that are slightly hooked. Many, including the blue-headed vireo, forage in a slower and more methodical manner than warblers, which are often in constant motion.

Blue-headed vireo, Pier 1

*Red-winged blackbird
(female), Pier 5*

RED-WINGED BLACKBIRD

(Agelaius phoeniceus)

On Pier 5, the Ample Hills Creamery ice-cream shop sits in a small kiosk-sized building, topped with something even better than a cherry—a rooftop mini-habitat of shrubs and grasses. I love this simple way of maximizing urban green space and providing more room for birds to rest and forage. Yet for two years, it seemed that house sparrows and starlings were the only ones stopping by. Finally, one summer, I noticed an interesting silhouette atop the shop—too large to be a house sparrow, too long to be a starling. As I approached, the field marks of this female red-winged blackbird came into view. The bird didn't seem to mind my presence, unlike the few males of this species that I had spotted on Pier 1 who, even as they towered high above me out of my camera's reach, felt compelled to take flight the second I focused my binoculars on them.

While male red-winged blackbirds are admired for their bright red and yellow shoulder patches—called epaulets—some females have these patches as well. They range in color from an almost indiscernible brown to a bright reddish-orange that is seen when the bird's wings are spread.

SAVANNAH SPARROW

(Passerculus sandwichensis)

t's spring, and something special has arrived on the lawns of Pier 1—a Savannah sparrow. Its white, thinly streaked breast stands out against the green grass. Another hops out from under a tree. Then another. Soon, I'm watching half a dozen Savannah sparrows forage together on Vale Lawn. When they've finally had their fill of insects, the birds disperse and rest on the branches of the surrounding trees.

The Savannah sparrow is named after Savannah, Georgia, where ornithologist Alexander Wilson coined the name in 1811 after he collected one of the first specimens of the species there. Savannahs often show a patch of yellow in the area between the eye and bill, a field mark that is helpful in distinguishing them from the similar-looking song sparrow (a year-round park resident).

Savannah sparrow, Pier 1

Common tern, Pier 4

COMMON TERN

(Sterna hirundo)

Common terns do not like chocolate chip cookies. How do I know? Well, all field guide accounts of this bird point to its diet of mostly fish, some mollusks and crustaceans, and an occasional insect. Being a chocolate lover, I found it amusing when I saw this common tern perched at Pier 2 with its tiny, reddish-orange feet nearly hidden by chocolate cookie crumbles. For some reason, the tern chose to perch on a post topped with someone's discarded snack. While a tern may look similar to a gull, it certainly has a more refined palate; any local gull would have partaken in the sweet afternoon treat.

In May, the common tern arrives from its wintering grounds—mostly along the coasts of Central and South America—to breed in New York City. A small colony sets up shop less than a mile from Brooklyn Bridge Park on a defunct pier section on Governor's Island. (Local naturalist Gabriel Willow discovered the colony and worked with NYC Audubon to protect it.) There, the birds assemble sparse nests of shell, sand, and stone right on the cement. Throughout spring and summer, the terns frequent Brooklyn Bridge Park, perching on exposed pier pilings and railings as they scan for fish.

Common tern, Pier 2

GREEN HERON

(Butorides virescens)

After leading a morning bird walk, I parted ways with my fellow bird-watchers as they headed up the Squibb Park bridge to Brooklyn Heights. (Much to the dismay of locals, this handy footbridge is currently closed due to bouncing more than intended, even for a hydraulic bridge.) It had been a slow birding day with sightings of the usual suspects—American robin, song sparrow, and common grackle. One person even asked where all the brightly colored birds were. If only the light had hit one of those common grackles just right, its iridescence would have wowed us all.

As I scanned the pilings before leaving Pier 1, a familiar Florida friend came into focus—a green heron. I quickly looked up toward Squibb Bridge and attempted a wild wave at the group, much too far up the bridge to see it. I wanted to yell, "Look! A colorful bird—here it is!" But I was too far away.

Two years later, my friend Bob and I were surprised by this green heron, perched high in a honey locust tree on Pier 1. We watched in awe as it edged along a branch in real-life slow motion.

Green heron, Pier 1

Cedar waxwing, Pier 1

CEDAR WAXWING

(Bombycilla cedrorum)

Sitting on a bench overlooking the Magical Knoll, shaded by the sturdy, draping strands of sumac branches above, I heard the high-pitched whistle of the cedar waxwing. Something about the sound struck me as different—it was more constant and urgent than the usual waxwing tones. And very loud. I stood up to get a better glance at the branch under which I'd been sitting. There, resting on that very branch, was a nest full of cedar waxwing chicks.

These birds get their name from what looks like a drip of red wax on the end of some birds' secondary (inner) flight feathers. Each drip is actually a small flat extension of the feather shaft that contains the pigment astaxanthin, the same one that makes shrimp and salmon pink.

Cedar waxwing flock, Pier 1

MAGNOLIA WARBLER

(Setophaga magnolia)

Behind the benches overlooking Harbor View Lawn, something catches my eye and quickly disappears. What did I just see? A quick-moving, bright-yellow bird—but this doesn't help much; many spring warblers fit this description. I look at the trees, watching for movement, but all I see are leaves waving gently from the breeze, not the bounce of a branch that I'm seeking. Someone passes by and asks what I'm looking at. "There was a warbler, a very colorful bird, here just a minute ago," I say, wishing I could share the sighting. A moment later, the bird reappears—and there's no mistaking this magnolia warbler.

The bold and bright colors of the magnolia warbler make it one of the easiest birds to notice during migration, when it makes a pit stop in the park to hunt for insects in the trees and shrubs. Only the male sports the striking facial markings—the female has a gray face (and much lighter streaks on its breast). In fall, after the summer breeding season, the male tones things down a bit and looks more like the female.

Magnolia warbler, Pier 1

Baltimore oriole, top of Granite Prospect

BALTIMORE ORIOLE

(Icterus galbula)

n early spring, the purple flowers of the paulownia trees atop Granite Prospect fill the air with a scent so sweet it stops park visitors in their tracks. As they breathe in the fragrance of the nectar-filled flutes—hanging eye-level with a Manhattan skyline and Brooklyn Bridge as backdrop—people often ask me what kind of tree it is. Then they head off to enjoy the park's views and perhaps stop for a refreshing drink. But for the Baltimore oriole, the tree itself serves as the perfect park drink stand.

One morning I watched this male as it held on to the stem of the purple flower, pierced the bloom's base, widened the hole by opening its bill, and used its tongue to drink the released nectar. This ability to open one's bill with force is an icterid, or blackbird, family specialization, made possible by an adaptation of the birds' skull muscles. By drinking nectar, this oriole obtained sugars that quickly converted to fat to fuel its migration journey (though it may not have had much farther to go—the Baltimore oriole nests in numerous locations around the city).

Baltimore oriole drinking nectar

COMMON YELLOWTHROAT

(Geothlypis trichas)

A masked male common yellowthroat peeks out from a dense shrub, giving me an eye-level view of North America's most frequently sighted wood warbler. "Welcome to Brooklyn," I whisper, as I imagine the many struggles this bird encountered on the trip from its wintering grounds in the southern United States or the tropics to here, just south of Times Square. There are endless possibilities: a scarcity of insect fuel at a migration stopover spot, a near-death encounter with a feral cat, a fierce, turbulent storm requiring an emergency landing. Could any creature be more tenacious? Its head suddenly retreats back into dense shrub cover as it continues the search for food. Through the leaves, I see it gobble down a bright green inchworm and silently cheer it on to find more. Fuel up, beautiful bird, and be safe during your stay in the city.

Common yellowthroat,
Pier 1 Dark Forest

Common yellowthroat, Pier 1

Black-and-white warbler, Pier 1 wooded paths

BLACK-AND-WHITE WARBLER

(Mniotilta varia)

n the Dark Forest, I find a low wooden fence post topped with moving shades of salt-and-pepper—a black-and-white warbler is investigating a crack. The bird's bill disappears as it pecks rapidly into the gap. A runner races by, just inches from the post. The bird doesn't flinch. Finally, it finds a prize—a large, reddish arachnid. I watch as it eats its leggy snack.

When it touches ground, a migrating warbler's main goal is to find food to fuel its journey. The park's mini-habitats are an important stopover in this search for sustenance. The birds are always on the hunt: A yellow warbler flits through the peaks of the park's oak and sweet gum trees searching for insects, a common yellowthroat hops through the shrubs for caterpillars, a northern waterthrush wades at the pond's edge picking up tiny mollusks.

The black-and-white warbler takes yet another approach; it examines tree limbs and scales trunks—and wooden fence posts—in search of bark-dwelling insects. This foraging style is akin to that of a creeper, which tends to scale upward, or of a nuthatch, which tends to scale downward. However, the black-and-white warbler scales in all directions, moving up, down, and around tree trunks and branches.

*Black-and-white warbler,
Pier 1 Dark Forest*

SWAMP SPARROW

(Melospiza georgiana)

n early spring 2015, the chill of winter was still holding a tight grip on the city, and the park was cold and still. The loud two-note trill of the swamp sparrow rang out from the pond at the north end of Pier 1. I scanned around the margin. I found the bird hopping around a partially submerged stump topped with dying weeds and a clump of grass. Its tiny feet gripped the bark as it scaled the stump, searching for insects hiding in the cracks.

The swamp sparrow—closely related to the park's year-round resident song sparrow—visits the park during migration in spring and fall. It forages in the leaf litter of the wooded paths and the Magical Knoll, where it's nearly camouflaged by the surrounding leaves. In spring and summer, the swamp sparrow breeds in marshes throughout the Northeast and Canada.

Swamp sparrow, Pier 1 north pond

Ovenbird, top of Granite Prospect

OVENBIRD

(Seiurus aurocapilla)

At the tail end of spring migration, I heard a warbler singing on Pier 1. I just wasn't sure *which* warbler it was. Like clockwork, the bird's call rang out loud and clear every few minutes. I could tell it was close—possibly right under my nose. But I chose to look up. After all, this was a bird. And birds fly.

As the serenade continued, I diligently scanned the trees above, baffled at the lack of bird clues—no moving shadows, no fluttering leaves, no tiny feet visible on a branch. After forty-five minutes, I noticed an ovenbird doing its chicken-like strut through the bushes below. (This warbler almost always forages on the ground.) Then it hit me—was *this* the bird I had been hearing? Without a peep, the ovenbird disappeared into the brush.

Once home, I went straight to my laptop to listen to the song of the ovenbird. Yes, I had in fact been duped by one of my favorite warblers. It was the ovenbird all along. Though slightly embarrassing, this experience turned out to be the best way to learn a birdcall. The next day, the sound of the ovenbird welcomed me as I arrived on Pier 1. I sat down on the path and listened intently to each and every note of the bird's tune. The ovenbird then peeked out from under the flowering shrubs and smiled for the camera.

YELLOW-RUMPED WARBLER

(Setophaga coronata)

Many bird names can be misleading (and often frustrating) to new birders. A red-throated loon has a red throat only during its breeding season in Canada and Alaska; in winter it's mostly gray and white. A female black-throated blue warbler doesn't have a black throat and it isn't blue; as with many warblers, the colorful plumage pattern is worn only by breeding males. The list goes on. But there's one bird that stays true to its namesake in all seasons and sexes—the yellow-rumped warbler. While the male does molt to a dull shade of brown in fall, its yellow rump always remains. But don't be duped—both the magnolia and Cape May warbler also have yellow rumps. Isn't birding grand?

*Yellow-rumped warbler,
Pier 1*

Yellow-rumped warbler (female), Pier 6

Chipping sparrow, Pier 1 Harbor View Lawn

CHIPPING SPARROW

(Spizella passerina)

There's a term some use to describe sparrows and other birds that are mostly brown—little brown jobs, or LBJs. Some birders take offense at this term, and with good reason; *LBJ* lumps together many species with distinct field marks and behaviors, worthy of study and admiration in their own right. Although I've never used the term—I actually hadn't heard it until I saw it in Audubon's tongue-in-cheek Dictionary for Birders—I've certainly had the sentiment that there was no way I could identify all the small brown birds.

It wasn't until I spent quality time with sparrows in Brooklyn Bridge Park that I began to fully appreciate their beauty and fall in love with them. Now I always look forward to their arrival spring and fall—October and November are actually the best months for sparrows in the city—when I get to try sorting out the sparrows on the lawns of Pier 1.

This chipping sparrow was a spring visitor, and its breeding plumage made it easy to identify. In fall, the bird's cap loses its reddish tint, its bright white supercilium (a stripe that runs above the eye) turns tan, and all its browns become duller, rendering it much more of a—dare I say—LBJ.

GREATER SCAUP

(Aythya marila)

t was early June and the diving ducks of winter were a distant memory. The frenzied pace of spring migration had come and gone, and it was time to get ready for my favorite city breeders—common tern, barn swallow, and gray catbird. But when I passed Pier 4 beach, this greater scaup—a bird of northeast winters—set my birding track on rewind. While most of these diving ducks migrate to their breeding grounds in northern Canada and Alaska in spring, each year several stragglers remain in the city.

As I observed the male scaup's glowing and slightly shifty eyes through the lens, I was oblivious to a camera crew setting up on the rocks just twenty feet from shore. When I finally finished photographing, I turned around to see the crew and a sharply dressed, long-locked male model holding a surfboard, waiting for the photo shoot to start. Perhaps they had been waiting for my shoot—with a slightly different attractive male model as subject—to end. As I left the beach, I looked back at the scaup, now sleeping on the shore. I wondered if the dapper duck would appear as a background dot in the magazine, which, according to posted signs, happened to be *GQ*.

Greater scaup, Pier 4 beach

Wood thrush, Pier 1

WOOD THRUSH

(Hylocichla mustelina)

As I scanned below the trees at the base of the Magical Knoll, I noticed something that didn't quite blend in with the leaf litter—a wood thrush. Its bright white breast, dotted with spots of the darkest brown, appeared in sharp contrast to the dead leaves behind it.

This small thrush is not as common a sight as in decades past; it's the poster child of the unfortunate decline in migratory songbirds. In the north its nests are frequently parasitized by the brown-headed cowbird (forest fragmentation has made it easier for cowbirds to locate wood thrush nests). And in the tropics the wood thrush has lost the majority of its preferred wintering habitat—primary forest—to deforestation. It now resorts to the small patches of forest that remain.

And so I cling to my few park sightings of the wood thrush, now colorful images burned into my brain. There was the one resting under the trees of the Magical Knoll, another fumbling with a berry in the Exploratory Marsh, and this one, foraging behind the benches along Bridge View Lawn.

YELLOW WARBLER

(Setophaga petechia)

Along the Brooklyn waterfront, in the darkness before dawn, the lights of lower Manhattan are the only hint of the city—no traffic on the street, no helicopters in the sky, no sound of sirens or honking horns. In the park at Pier 6, the northern waterthrush bobs near a puddle in the water playground, its streaked white breast glowing in the still dim light. Just across the path, warblers begin to flutter; a black-and-white and a yellowthroat squabble on a branch, with the former emerging victorious.

Their day has begun and it is now time to hunt for the choicest of caterpillars and spiders and an occasional berry (protein-packed snacks are most important to fuel the birds' migration flight). For the yellow warbler, aphids are on the early-morning menu in the nearby Exploratory Marsh.

The yellow warbler is a common park migrant in spring and fall. Its favorite foraging spots include the catalpas and paulownia trees on top of Granite Prospect, just under the canopy of the Dark Forest, and in the treetops on Bridge View Lawn. This colorful songbird breeds in Jamaica Bay Wildlife Refuge in Queens as well as farther north and into Canada. In fall, it migrates to its wintering grounds in Central and South America, often flying nonstop over the Gulf of Mexico (a feat also accomplished by the much smaller ruby-throated hummingbird).

Yellow warbler, Pier 6 Exploratory Marsh

Warbling vireo, Pier 1

WARBLING VIREO

(Vireo gilvus)

The warbling vireo is aptly named after the cheerful song of the male—a pretty, varied warble of notes lasting about three seconds. But when I came across the bird's name in my first field guide, I thought it was a joke. I couldn't yet identify most warblers and I couldn't yet identify most vireos. And now I could soon be faced with the task of identifying a *warbling vireo*. The plain-faced look and gray color of the bird didn't make the situation seem any easier. I met the bird a handful of times in Florida—always identified and pointed out by field trip leaders. I wondered if I could identify this bird on my own.

One day while birding in Brooklyn Bridge Park, a vireo poked its head out from a tree just south of Pier 2. Much to my surprise, there was no mistaking it—the bird's plainly marked face and thick bill screamed "warbling vireo." In summer 2015, a pair of these birds foraged about Pier 1 for nearly two months. Each day I watched intently as they flew perfectly in sync, with just inches between them. The sunlight reflected off their light plumage as they wove gracefully through the trees and out of sight—the male warbling all the way.

The warbling vireo nests near humans in parks and neighborhoods around the city. Though I have yet to find a nest in the park, their long summer stay points to good things to come. In fall, the vireo migrates to its wintering grounds in western Mexico and the northern half of Central America.

EASTERN KINGBIRD

(Tyrannus tyrannus)

t was a gorgeous but sweltering day in the middle of August. I was out in the park by 6:30 AM and was greeted by a yellow warbler on Pier 2 and barn swallows perched on the railing below the Pier 2 sports courts. After some great birding on Pier 1 the day before, I was expecting a repeat or better. But things were slow. So slow that I spent about an hour photographing a chocolate cupcake that was "perched" on a low fence post at Vale Lawn. But this wasn't just any cupcake. The beautifully decorated treat, complete with candle and bright-colored candy toppings, was moving, its entire lower half covered in ants. Having recently watched a documentary on E. O. Wilson and his work with ants, I was fascinated.

Figuring all of the bird action had passed for the morning (it was now 11 AM), I decided to switch lenses to get some landscape shots. Within minutes, I noticed a large bird with black upper parts hiding in the top of a tree—an eastern kingbird. I raced to a bench and switched back to my "bird" lens, but by the time I returned, the bird was high in a catalpa and out of photo range. I waited patiently in somewhat oppressive direct sunlight, watching the bird and trying to keep tabs on it. After forty-five minutes, the kingbird finally flew to a shrub and perched perfectly on top.

Ant cupcake, Pier 1 Vale Lawn

*Eastern kingbird eating
magnolia seeds, Pier 1*

Chestnut-sided warbler (fall), Pier 1 wooded paths

CHESTNUT-SIDED WARBLER

(Setophaga pensylvanica)

My black-tea buzz hadn't kicked in yet, but there was something waiting to give me a jolt right at Pier 6. Just as I set foot in the park, a male chestnut-sided warbler looked down at me from among the dangling catkins of a tall, thin oak tree. I tilted my head back as far as I could—risking that neck pain that birders call "warbler neck"—and marveled at its prominently patterned spring breeding plumage.

In its fall (nonbreeding) plumage, this warbler looks very different from the one I saw that day in May: It has a glowing lemon-lime crown, a bold white eye-ring, and pale gray cheeks and neck. Just a hint of its namesake still remains in faded fall flanks, and its black mustache won't grow back until breeding season.

Chestnut-sided warbler (spring), Pier 6

EASTERN WOOD-PEWEE

(Contopus virens)

A mourning cloak (*Nymphalis antiopa*), an early-spring butterfly, moved leisurely above the meadow of the Magical Knoll in search of the perfect flower. The light yellow borders of its wings appeared bright white in the reflective midday sun. Then suddenly—poof! It was gone. It took me a second to process what just happened: An eastern wood-pewee had snatched the butterfly out of thin air.

During migration, the eastern wood-pewee catches insects in an often dizzying aerial display above the Magical Knoll. It loops, dips, and dives through the air, as if tracing the track of an invisible roller coaster. Between flights, it may perch on the stem of a hanging catalpa pod that offers a perfect view of the many beetles, wasps—and butterflies—that fly below.

Eastern wood-pewee, top of Granite Prospect

*Empidonax flycatcher, Pier 6
Exploratory Marsh*

EMPIDONAX SP.

During migration, flycatchers are a common sight on Pier 1, where they perch low in trees and shrubs surrounding the Magical Knoll, frequently dipping and hovering beside the brush to snatch low-flying insects. But this *Empidonax* flycatcher—"empid" in birder speak— caught me by surprise one afternoon in the Exploratory Marsh at Pier 6. Things were unusually quiet for a spring day, save for the babbling of the marsh's water pump. With no sign of a single bird, I leaned forward slightly against the marsh's metal railing to peek around a shrub when my eyes fell upon this flycatcher, perched perfectly still, looking down at me. Ever since, that round-the-shrub peek has become a birding ritual, though it has yet to reveal the beauty it did on that quiet day in May.

This flycatcher is likely a yellow-bellied (*Empidonax flaviventris*), but since I didn't get a look at the front of the bird or hear its call, I can't positively identify it as such. I have to settle for an unscientific description—an extremely cute and compact empid.

BLACKPOLL WARBLER

(Setophaga striata)

When I was a little girl I used to sit at our family's upright piano, striking the highest key and giggling at its shrill sound. I did much the same to prepare to meet the blackpoll warbler, but this time, I was hitting the play button on the species' audio track on Cornell Lab of Ornithology's All About Birds Web site. The blackpoll, champion migrant among warblers, has one of the highest-pitched calls of any bird, with a frequency more than double that of the highest key on an eighty-eight-key piano (9,000 hertz versus 4,000 hertz). I was determined to commit this call to memory so I wouldn't miss a chance to add the species to my life list. My sound study paid off when I heard an almost inaudible *tsit-tsit-tsit* during a trip to Cape May, New Jersey, and located the blackpoll warbler.

Just four days later, guess what I finally saw in Brooklyn Bridge Park? A male blackpoll—perched tall, still, and silent in a small oak on Vale Lawn. Males show a black cap in breeding season; in fall, they molt into their basic olive-green plumage and look similar to females and immatures.

Blackpoll warbler, Pier 1

Blackpoll warbler, Pier 1

Louisiana waterthrush, Pier 1 Long Pond

LOUISIANA WATERTHRUSH

(Parkesia motacilla)

The Louisiana waterthrush looks nearly identical to the northern waterthrush of the same genus. Both birds visit the Pier 1 Long Pond during migration and both constantly bob their backsides. But there are many field marks and behaviors that make it possible to tell them apart. The streaking on the Louisiana's breast is lighter, not as dense, and does not continue all the way up the throat. Its underparts are usually whitish compared to yellowish on the northern. Leg color provides yet another physical clue. In spring, the Louisiana's legs are usually brighter pink, while the northern's are dull pink or gray.

Even armed with knowledge of the physical differences between these two waterthrush species, identification can sometimes be tricky, especially with juvenile birds. Other factors such as breeding habitat, timing of migration, and differences in song can also help. For example, during spring migration, the Louisiana waterthrush is the first of the two to arrive in the city.

WOOD DUCK

(Aix sponsa)

Just south of Pier 1, dozens of exposed pier pilings jut from the water in slightly curved rows, foregrounding the lower Manhattan skyline behind. On summer evenings, photographers gather here just as the golden light of sunset combines with the gradual illumination of skyscraper lights.

The pilings are a popular perch and resting spot for cormorants and gadwalls, American black ducks, and mallards. On winter nights no piling is left vacant as hundreds of gulls—mostly ring-billed and some herring—occupy each and every one. Unlucky latecomers must roost in the water below.

While it's tempting to quickly eyeball the pilings for a count of these common birds,

after spotting a green heron there in 2013, I vowed to always check each and every one. It wasn't until two years later that this ritual finally paid off. As I scanned, I passed by an arresting set of colors and my eyes landed on a treasure—a male wood duck. Its blue, purple, red, and green plumage decorated the top of the otherwise drab and faded wooden support. The bird stayed on the same piling for several hours and was there when I left the park.

Wood duck, Pier 1 pilings

Wood duck (female), Pier 1 pilings

Yellow-throated vireo, Pier 1 Dark Forest

YELLOW-THROATED VIREO

(Vireo flavifrons)

Once I reached my goal of spotting 100 species in the park, my quest didn't stop, but finding new species became more of a surprise than a hope. So I was shocked the following spring when, within a week, I added five new species to my park list.

The wooded paths of Pier 1 were teeming with park visitors. It had been a brutal winter, and it seemed as if every New Yorker had burst outside in unison to soak up the warmth of the spring sun. (Not everyone enjoys watching ducks for hours in subzero temperatures.)

Just as I was doubting that any good birds would join the crowd, this yellow-throated vireo flew in from the east and perched in a tree right in front of me. Its bright yellow throat and breast illuminated a shady spot under the tree canopy. I admired the bird's thick bill, olive-tinged head, and yellow goggles for a brief moment before it flew off. There is a plus to such a brief bird sighting; I captured just two photos of the bird and didn't have to sift through hundreds to find the good ones.

MARSH WREN

(Cistothorus palustris)

Marsh wrens are usually heard before they are seen, if they ever get seen at all. But this one never even made a peep. I noticed a brownish bird right when I arrived at the Long Pond on Pier 1 and was in disbelief when I focused on its even darker brown cap, a diagnostic mark for the marsh wren. Great, I had identified it. But this was no time to savor the moment—I had to act fast before this notoriously secretive bird took cover.

As I sat down to appear less threatening, the bird perched head first on a vertical branch, nearly blending in with the dead leaves on the slope behind it. My body grew tense as the wren hopped down to the water's edge. What would it do next?

It flew over my head to the other side of the walking bridge on which I was sitting. With the bird now behind me, I held tightly on to my camera and spun around on my behind, not caring about the harsh metal grate below me. (This was fun—I never knew I had it in me. I felt like a pro, determined to get the shot.) As I turned around, I held my breath when I saw the bird in the most prized of all poses—perched straddling two branches, with a leg on each.

Marsh wren, Pier 1

White-crowned sparrow,
Pier 1 Little Shrub Stand

WHITE-CROWNED SPARROW

(Zonotrichia leucophrys)

When submitting a checklist to eBird—my preferred online tool to record and view bird sightings—you are presented with a list of all species that can be reasonably expected at that location that time of year. Just looking at the checklist is a great way to learn the birds of your area and builds anticipation about potential sightings. The white-crowned sparrow was always on my eBird checklist for the park during migration, yet, after two years, I had yet to spot it there. (Potential sightings are not necessarily common sightings.)

The bird's preferred habitat of bare ground and grasses meant it might drop by eventually. But it wasn't until 2015, while I was sitting on a bench next to Vale Lawn, that the white-crowned sparrow hopped up to meet me, clocking in as my 118th park species. It stayed in the park for a week, often edging out from under the shrubs lining the paths of Pier 1.

While western populations of this species are year-round residents and don't migrate, most white-crowned sparrows migrate north to breed in northern Canada and Alaska.

HOUSE FINCH

(Haemorhous mexicanus)

One afternoon, Connor and I continued our long park stroll on to Vinegar Hill, a neighborhood that lies just north of the park boundary. As we passed through the cobblestone streets lined with industrial warehouses and marveled at a large power plant that looked like something out of a sci-fi movie, we heard a bird singing an assertive melody. It was a house finch, perched on the roof of a brick building. As it sang, I spoke to it: "You're so gorgeous. Why don't you come visit me in the park?" Though a common year-round city resident, the house finch mainly frequents parks and backyards with bird feeders.

The next day, when I arrived at the Long Pond, it was devoid of bird activity. I thought back to some of my bird encounters there: the Lincoln's sparrow that appeared in the dead shrub for a split second, the northern waterthrush that captivated me for hours with its bobbing and wading, and, just a few days earlier, the secretive and rarely seen mourning warbler that showed up in a shrub that lined the pond. Then I heard something behind me. It was birdsong—elegant, assertive birdsong—that sounded familiar. I turned around. There, plucking berries from a tree, was this house finch.

House finch, Pier 1 Long Pond

Young Birds on the Waterfront

Each spring, nearly a dozen bird species go to work staking out nesting locations and gathering nest materials in Brooklyn Bridge Park. American robins collect twigs from leaf litter, barn swallows scrape mud from the edges of the park's ponds, and house sparrows peel thin wood fibers from tree supports. By midsummer, Pier 1 becomes a Brooklyn bird nursery as fledglings dot the lawns, their mouths wide open, begging to be fed. In the water between the park's piers, female mallards circumspectly lead their ducklings to the water's edge to forage. In more hidden locations, dove fledglings, including pigeons, place their beaks inside that of their parents and feed on crop milk. While many birds have a crop for food storage, only doves are known to produce such milk for their young.

American robin and hatchlings, Pier 1

Barn swallow feeding fledgling, Pier 6 Governors Island Ferry Landing

Common grackle fledgling, Pier 3

Gray catbird fledgling, Pier 1

European starling fledgling, Pier 1 entrance

Northern mockingbird fledgling, Pier 3

American robin
fledgling, Pier 1

House sparrows, Pier 5

Female mallard and ducklings, Pier 4 beach

European starling
fledgling, Pier 1

Canada geese with gosling, Pier 4 beach

Gray catbird, Pier 1, Long Pond

SUMMER

Hot and humid city summers are a busy time for the many birds that nest and raise their young in the park. It's also the best time to observe the many behaviors of these birds on their breeding grounds. On a summer stroll through the park, one can witness courtship, nest building, mating, feeding of young, competitive and defensive scuffles, and more.

By late July, "fall" migration begins, as birds that have raised their young elsewhere in the city and farther north begin to pass through the park. And as summer ends, breeding season draws to a close. Young barn swallows have fledged and are ready to depart. They line up on the railings of Pier 2 with their parents, prepping for the flight. Only a single common tern remains; it squabbles with a gull over a piling to perch on. The gull wins and I won't see another tern until spring.

AMERICAN BLACK DUCK

(Anas rubripes)

The park was quiet for a weekend day. The morning cloud cover and wind caused a slight chill that hinted at the first day of fall, just a few days away. Two American black ducks swam out from the spiral pool, nearly blending in with the dark water below. As they headed south under Pier 2, the sound of a single bouncing basketball echoed from the sports courts.

The spiral pool is a popular resting and foraging spot for both American black ducks and mallards. At high tide, the ducks pluck algae from the pool floor with their heads submerged and their tails and behinds poking straight up from the water. At low tide, starlings join the ducks in picking the algae off the exposed cement.

The American black duck is actually dark brown, but it can look black from a distance. It looks very similar to the female mallard (and to the male mallard in nonbreeding/eclipse plumage). One way to pick out black ducks from mallards is by looking at their speculum—a blue patch of flight feathers that is bordered in black. On mallards, the speculum is also blue but is bordered in bright white. Sometimes, the identification is not as clear-cut; black ducks are one of the many waterfowl with which mallards hybridize. Many of these mallard × American black duck hybrids have a combination of both species' physical features. For example, such a hybrid may have dark body plumage (like the black duck) and green head feathers (like the mallard).

American black duck, Pier 4

Song sparrow, Pier 1

SONG SPARROW

(Melospiza melodia)

One rainy day in June, a chocolate craving struck. Connor and I grabbed our umbrellas and headed to the park on our way to Jacques Torres Chocolate in DUMBO. At Pier 1, a song sparrow landed on a low fence pole. We stopped to watch through the raindrops (we had left our binoculars at home). The bird looked around and lifted its tail feathers; another song sparrow landed behind it. The two mated and the female flew off.

Not bad for rainy-day birding; the excitement almost made me forget about my chocolate craving.

About a month later, in the same area, a song sparrow edged along a tree branch. As I brought the bird into focus, I noticed two white tufts on either side of its head—this was a fledgling! I wondered if this bird was the result of that rainy-day romance in June. An adult flew in with a food delivery and fed its young a small caterpillar.

The song sparrow is a year-round park resident. Its song—a few evenly spaced notes followed by a trill—can be heard ringing out over Pier 1 in spring. On winter's snowiest days, the song sparrow often feasts on the seeds of the little bluestem, a prairie grass found on the Pier 3 uplands.

LAUGHING GULL

(Leucophaeus atricilla)

Years before I ever donned a pair of binoculars, I was struck by the beauty of the laughing gull. I remember snapping photos of the bird with my little point-and-shoot camera as it stood outside my Florida apartment. The white arcs around its eyes looked painted on; its dark red bill was striking.

Laughing gulls arrive in Brooklyn beginning in April and during the summer when they can be seen flying around the park and resting in the water and on the Pier 1 pilings. Only breeding adults have the striking black heads, but the bird's size and flight style provide identification clues in its various other plumages; it is slightly smaller than the park's common ring-billed gull, with thinner wings and more agile, graceful flight. The laughing gull gets its common name from one of its calls—a series of descending notes that sound like frenetic laughter.

Laughing gull, Pier 5

Gray catbird, Pier 1 Dark Forest

GRAY CATBIRD

(Dumetella carolinensis)

When I lived in Florida, I would see the gray catbird on its wintering grounds, but not frequently enough to study it in the field. My first spring back in Brooklyn, I was thrilled to see a catbird arrive the first week in May. Little did I know that this bird—and the many other catbirds that would follow in the coming weeks—had a surprise in store. They had chosen Brooklyn Bridge Park as a breeding ground and would provide me with endless hours of observation as they courted and paired up, constructed nests, and raised their young just steps from my apartment.

In early spring, adult catbirds attract mates by singing loudly while perched under the low canopy of Pier 1. Engaged in courtship, they dart in flight across the wooded paths, as if in a race.

By mid-July, the high-pitched begging calls of catbird fledglings are the background soundtrack for a stroll through the Dark Forest. First, one hungry fledgling will call, then another. Soon, the entire Dark Forest comes alive with a crescendo of young catbird calls as the adults scramble to make on-time deliveries of berries and insects.

AMERICAN CROW

(Corvus brachyrhynchos)

t wasn't until I captured this photo that I realized I had a serious crow identification problem. I had spent my time studying warblers and sparrows, not realizing there was something simple yet important I needed to know about crows. I had always assumed that if I ever managed to get a decent photo of a crow (a difficult task, as they don't stay put for long in the park, and capturing the detail in their dark plumage is impossible without perfect light), I could simply use the photo to identify it by its physical features. Turns out I was wrong. While the American crow is larger than the fish crow, differences in age and sex account for plenty of overlap between the two species in regard to body size, bill shape, relative primary feather length, and more. There is only one way to reliably distinguish them: by their call.

While I did hear this crow's call, I didn't realize at the time how important it was to remember it. So I started paying careful attention to every crow I came across, comparing the calls I heard with those on birding Web sites. It was pretty easy to tell the difference by sound: the American crow a drawn-out *caw-caw*, the fish crow a weak *uh-oh*. I should have taken the time to learn the calls long ago. I had been birding for five years, but only knew the calls of birds I saw frequently or target birds I was hoping to find for the first time.

So you may be wondering—how do I know this bird is an American crow? More than a dozen times, I've closed my eyes and traveled back in time to the day when I took this photo. I picture myself looking at birds on Bridge View Lawn when I hear a crow call. I walk toward the source of the sound to find the bird watching me from a tree. It calls again. Every time I do this visualization, I hear the *caw-caw* of the American crow.

*American crow, Pier 1
Harbor View Lawn*

Brown-headed cowbirds, Pier 1

BROWN-HEADED COWBIRD

(Molothrus ater)

The brown-headed cowbird is one of the least-loved birds in North America. Brood parasites, cowbirds lay their eggs in the nests of other species, usually ejecting one of the host's eggs and damaging, or even eating, the others. When a cowbird egg hatches, the bird is fed and raised by the host species, which, heartbreakingly, is often much smaller than the young cowbird.

As I walked on the path near Harbor View Lawn, I heard strange and repetitive gargling noises. I looked up to find two cowbirds engaged in courtship. Facing the female, the male cowbird spread its wings, puffed out its feathers, and extended its neck in an eerie, Dracula-esque fashion. It ended this sequence with a bow to the female, who looked pretty uninterested.

It wasn't until the birds flew away that a feeling of dread hit me. Assuming the male got the girl, where would she be laying her eggs? I worried about the park's nesting gray catbirds, American robins, and especially the much smaller song sparrows but reminded myself that nature isn't always kind. And while I did feel lucky to have witnessed the macabre beauty of the cowbird courtship display, I hoped I wouldn't have to witness a host bird feeding a cowbird fledgling (which, luckily, I haven't yet).

RED-EYED VIREO

(Vireo olivaceus)

The park had been good and birdy one late-summer morning. At dawn, the south end's Exploratory Marsh was teeming with several each of black-and white warbler, common yellowthroat, yellow warbler, and northern waterthrush. On the north end, at Pier 1, while two Carolina wrens were loudly caroling, a Wilson's warbler and a yellow warbler gleamed in silence against the dark leaves of the pin oaks in which they foraged. (Warblers sing little if at all during fall migration; they are not trying to attract a mate.) A red-eyed vireo peered out from under some leaves as several scarlet tanagers flew low over Vale Lawn. The heat of noon arrived all too quickly; it had been a great birding day, though nothing had come in close enough for a photo.

My obsession was evident as I headed back out in late afternoon. Had I lost my mind? Possibly. But something special awaited me at the Little Shrub Stand on Pier 1. This small but dense mini-patch has a combination of plantings that has proven to be a bird magnet: a pair of American hollies, a Kentucky coffee tree, a sweet bay magnolia, and dogwood shrubs, flanked by a honey locust and plane tree. When I passed the stand before sunset, the leaves of the small sweet bay magnolia were moving. Though I couldn't see a bird, past treasures of this very tree—scarlet tanager and red-eyed vireo—whirled in my head. As I edged forward, this red-eyed vireo moved out into view to feed on magnolia seeds.

Red-eyed vireo, Pier 1 Little Shrub Stand

Black-crowned night-heron, Pier 1

BLACK-CROWNED NIGHT-HERON

(Nycticorax nycticorax)

Hoping to track down a small bird that flew into some trees on Pier 1, I stopped and poked my head in and around the tree trunks. As I looked up, my eyes fell upon two giant, yellow, prehistoric-looking feet, gripping a branch. Those feet happened to be attached to this black-crowned night-heron, a species that stands about two feet tall. Talk about a consolation prize.

The bird continued to use this location as its daytime roost throughout the summer but decided to pass on the US Open tennis large-screen viewing event on Harbor View Lawn for two weeks in September. A few days after the event was over, I was relieved to see the bird had returned; it continued to give me a rush when I visited it during the remaining days of summer.

And to think this sighting all started by my following a little brown bird (likely a house sparrow). Even though most small brown city birds don't turn out to be something rare, you never know where they might lead you. This coincidental discovery of the night-heron fed my obsession about not missing a bird. As a result, I can often be seen heading home and then turning around as if I've forgotten something in the park. What's actually happening is one of two things: Either I just saw a great bird fly by and I follow it, or I'm feeling anxious about not having checked a certain location and decide to head back just in case.

Actually, there is a third possibility—I might have made a last-minute decision to head to nearby Jacques Torres for hot (or frozen hot) chocolate.

FISH CROW

(Corvus ossifragus)

While the *uh-oh* calls of the fish crow sound harmless, they can be quite ominous to other birds within earshot. The crows utter these calls from their favorite hangout in the 'hood, the beer distribution center adjacent to the south end of the park. From the facility's rooftop, the crows keep a close eye on the barn swallows that nest under Pier 6, waiting for a chance to swoop in on an unguarded nest and grab a tasty little barn swallow meal. While this crow is named for its penchant for fish, I've never seen a fish crow eat a fish in Brooklyn Bridge Park, or anywhere else, for that matter. And luckily, I've never seen one eat a bird. I have seen many a crow flying in the distance with some unidentified creature dangling from its mouth. At this sight I, too, say *uh-oh* and try to convince myself it's a house sparrow (a nonnative "pest" species) in the corvid's grip.

Fish crow, Pier 5

Great egret, Pier 2

GREAT EGRET

(Ardea alba)

One summer Sunday, my friend Bob and I met for our weekly birding and brunch. Not expecting any shocking sightings in the middle of summer (those tend to happen during spring and fall migration), we strolled leisurely through the park while catching up on the events of the week. Still, my now-instinctual peripheral visual and auditory bird monitoring system could not be turned off. Common terns squeaked aggressively as they competed for small fish. Barn swallows filled the air with a bubblier sound that was at the same time subtle and serene. A female mallard and her ducklings, still sporting tufts of fluffy down, swam alongside the rocks at the foot of Pier 3.

As we approached Pier 2, a large white apparition ascended from behind the rocks of the spiral pool. After a brief moment of disbelief, I realized this great egret was indeed flying right in front of us. We watched, mesmerized, as the bird circled over Pier 2 and headed south.

While a great egret is not rare for the area in summer—it can easily be sighted in Jamaica Bay Wildlife Refuge and other area wetlands—it is uncommon to see one in the park, which only offers a tiny patch of the bird's preferred shallow-water wetland habitat. This egret likely stopped by the park to scope out the urban food menu and quickly decided to go elsewhere.

CAROLINA WREN

(Thryothorus ludovicianus)

The wooded paths of the Dark Forest were alive with the sounds of warblers and wrens. Shadowy birds darted through the darkness, passing briefly through thin beams of the early-morning light breaking through the canopy. Two ovenbirds flitted between branches, uprooted from their ground foraging by two Carolina wrens now hopping about. After spending the better part of the day tracking the wrens' loud song—*teakettle-teakettle-teakettle*—I was fortunate that they chose to end the day on the shrub stand at the east end of Harbor View Lawn. Here, at sunset, the two perched on the tightly packed holly branches, and occasionally one or the other would fly into focus on the taller trees beside them.

Carolina wren, Pier 1 Little Shrub Stand

*Willow flycatcher,
Pier 1 Magical Knoll*

WILLOW FLYCATCHER

(Empidonax traillii)

have a tough time distinguishing flycatchers in the genus *Empidonax* by sight. But so does everyone else. These birds, fifteen species in all, and five normally occurring in the northeast United States, look virtually identical. While there are certain field marks that provide clues, some of these species can reliably be identified only by sound.

One week in August, one of these empids was catching insects midair on the Magical Knoll. It did sing (not really a song, more like a high-pitched buzzy whistle that is really quite cute), but I hadn't yet delved into the study of flycatcher ditties. It was time. When I heard the bird vocalize, I repeated the sound I was hearing—*weet-woo*—several times aloud. Luckily no one was around, though that wouldn't have stopped me. Then I jotted the funny phrase down in my naturalist notebook.

When I got home and entered my eBird checklist, I saw that the willow flycatcher was the most-to-be-expected empid in Brooklyn for this time of year. Maybe if I listened to the sound of a willow flycatcher, I would get a positive ID. I had my doubts. But when I hit the play button on the Cornell Lab of Ornithology's sound file, and I heard those two magic syllables—*weet-woo*! I also learned that the widely accepted mnemonic for the willow flycatcher's song is actually *fitz-bew*, but in my opinion, it doesn't really sound like that. This is often the case with birdsong mnemonics.

BLUE-WINGED WARBLER

(Vermivora cyanoptera)

The Dark Forest was the place to be one hot and humid week in August. Its canopy provided the perfect shade cover for both me and an adorable squirrel nibbling on an acorn. The loud *chink* calls of a waterthrush rang up the path, nearly masking the cheery notes of what sounded like several other songbirds. I walked slowly and nonchalantly (this feigned insouciance sometimes keeps birds from darting away at the sight of me) toward the action, until I saw not one, but two northern waterthrushes, an American redstart, a black-and-white warbler, and this blue-winged warbler, my 127th park species.

The blue-winged warbler often hangs upside down from tree limbs checking clusters of dead leaves for insects. And at this time in August, the Dark Forest offered an array of these leafy goldmines. One of them turned up quite the treasure—this bird scored a large and, I guess, tasty insect. Long after the bird departed the park (it only stayed a few days), several black-and-whites—also acrobatic warblers—continued to forage in these clusters, lapping up large doses of hidden protein well into fall.

Blue-winged warbler, Pier 1 Dark Forest

Budgerigar, Pier 3 uplands

BUDGERIGAR

(Melopsittacus undulatus)

O nce in a while an escaped pet bird shows up in the park. In summer 2015, two budgerigars seemed right at home as they foraged and flew alongside house sparrows at the Pier 3 uplands and Pier 1. Birders don't count the budgerigar on their lists; only birds with established North American breeding populations are valid. Still, these beautiful parakeets entertained many park visitors (and me) that summer. The birds were gone before the first chill of fall. I hope they found an open window and a warm home.

Brooklyn Bridge

FALL

It's the first week of fall, and along the wooded paths of Pier 1, the leaves rustle in the wind as if on cue. A high *seet* sounds; the first white-throated sparrow has arrived. The canopy of the Dark Forest starts to thin.

Migration is at its peak. Dozens of palm warblers descend on the park's lawns. Some flit around in the marshes alongside common yellowthroats. Thrushes peer out from the edges of the wooded paths. On Granite Prospect, warblers gather insects from the undersides of the now-fading catalpa leaves. A pair of bright brown thrashers forages on Vale Lawn.

A few weeks into fall, migrating sparrows join palm warblers on the lawns; others choose the Pier 3 uplands as a pit stop. With the urgency of breeding season behind them, birds take their time as they make their way south, with some spending several weeks or more in the park.

In the water between the piers, the gadwall—the park's most common wintering dabbling duck—now joins mallards and American black ducks. Its arrival hints at the more ornate diving ducks of winter; as the temperatures drop, my attention turns to the water.

Ring-billed gull, Pier 1

RING-BILLED GULL

(Larus delawarensis)

On a slow birding day in fall, I came across this ring-billed gull perched on a Pier 1 railing. The Brooklyn Bridge was a perfect backdrop, but there was one problem—to get the right angle I had to sit smack dab in the middle of a pedestrian path. The park wasn't packed on this late-November morning, so at the risk of looking like I was staging some sort of sit-in, I took a seat on the cement.

I got some strange looks, but it was worth it. The gull rewarded my admiration and efforts by belting out its long call with what seemed like a classic Brooklyn accent.

HERMIT THRUSH

(Catharus guttatus)

n spring, most migrating birds pass through the park quickly, fueling up for just a day or two before continuing on to their breeding grounds to claim a top nesting spot. Many birds on my park list have made an appearance for one day only, foraging high atop the trees and out of my camera's reach. In fall, many migrants take their time as they head south, with some spending a few days to a couple of weeks in Brooklyn Bridge Park. Warblers with long fall stays include the northern waterthrush, ovenbird, and yellow-rumped warbler. Of the three common park *Catharus* thrushes—Swainson's thrush, veery, and hermit thrush—the latter hangs around longest. It can be found foraging along Pier 1's wooded paths throughout October and November.

The hermit is the only *Catharus* thrush that winters in the United States. While most choose the southern states as their wintering grounds—with fewer migrating farther south to Central America—some stay in the north throughout winter.

Hermit thrush, Pier 6 Exploratory Marsh

Ruby-crowned kinglet, Pier 4

RUBY-CROWNED KINGLET

(Regulus calendula)

Ruby-crowned kinglets are one of the easiest migrants to spot in the park during spring and fall. Tiny birds in constant motion, they flit about, often hovering near the underside of large leaves to glean insects. Males have red crown feathers that they display when excited or agitated.

I took this photo one October afternoon when the small pines along the paths of Pier 4 were teeming with kinglets. I counted fifty in the park that day and there were likely even more.

Ruby-crowned kinglet, Pier 1

NORTHERN FLICKER

(Colaptes auratus)

After a slow morning of birding, I stopped by the Long Pond (a large pond in the middle of Pier 1's freshwater gardens that run along the main park path) to find a ruby-crowned kinglet flitting among the white aster flowers. As I looked out over the pond, I admired the fall scene—the long limbs of the red-leaved sumac extended almost vertically out over the duckweed-covered water. A robin flew in and began to bathe, puffing its feathers so far out it reminded me of a large hen. A male towhee came next and joined in bath time; things were really picking up. Just when I thought it couldn't get any better, a male northern flicker edged out from under a plant and began drinking calmly at the pond's edge. I watched as it dipped the tip of its bill into the water and gently tilted its head back to drink.

Fall is the best time to observe flickers in the park. Their yellow-shafted feathers and white-rump patches flash brightly as they fly across Harbor View Lawn. Here, they can also be seen digging for ants and beetles in the dirt at the bases of the trees. Unlike most woodpeckers, flickers usually feed on the ground, collecting insects with their long, barbed tongue. While many flickers are year-round city residents, the temporary visits of the park flickers during spring and fall suggest they are migratory. (Little is currently known about the migratory patterns of these birds.) In the western United States, flickers have red-shafted feathers; overlapping populations of the two color morphs often hybridize.

Northern flicker, Pier 1 Long Pond

American redstart (female), Pier 1

AMERICAN REDSTART

(Setophaga ruticilla)

This tiny five-inch female redstart swept along the branches of the young trees lining the Harbor View Lawn steps. It was difficult to track as it wove in and out of what seemed like every leaf, searching for insects. In expected warbler fashion, it would come into view for a second or two and then quickly disappear. Just down the steps of the path, I noticed a fluttery movement—it was the redstart doing a bit of fly-catching, just inches above the steps. As it hovered and caught flies, it fanned its tail, revealing stunning patches of bright yellow against dark gray.

Both male and female redstarts fan their tails often, especially when they move among branches. When flashed, the tail's bright colors—even brighter on the male, which is black and orange—are thought to startle insects and cause them to try to flee, enabling the redstarts to snatch an easy meal.

American redstart, Pier 1

LEAST FLYCATCHER

(Empidonax minimus)

The least flycatcher has a favorite park perch—a low-lying dead branch of a shrub on the Magical Knoll. I watch it from above on Granite Prospect; every half minute or so, it sallies out over the clearing to quickly snatch an insect. Within a few seconds, it's back at its perch, eyeing the knoll for the next passing gnat.

The smallest *Empidonax* in eastern North America, the least flycatcher uses the park as a stopover during migration. In spring, it's en route to its breeding grounds farther north—places such as the Adirondacks or the parklands of Manitoba, Saskatchewan, and Alberta. In fall, it's heading south to spend the winter in Mexico or Central America.

Least flycatcher, Pier 1 Magical Knoll

Swainson's thrush, Pier 1 Magical Knoll

SWAINSON'S THRUSH

(Catharus ustulatus)

The Swainson's thrush is one of many birds described in field guides and on Web sites as "more often heard than seen." This information can be discouraging to new birders. I remember thinking I would never lay eyes on an ovenbird, an American bittern, a Virginia rail—species all described in this way. But once you delve into birding, you discover it is possible to see most of these birds, even if only occasionally, as long as you watch for them in the right habitat at the right time of year. (Though something like the black rail can be next to impossible to spot. If you see one, please give me a call.)

While the Swainson's thrush is in fact secretive, it's not difficult to catch a quick glimpse of it during migration in Brooklyn Bridge Park. It usually happens like this—through my binoculars, I see the bird foraging on the ground, far up the path in the Dark Forest. As I approach, our eyes meet, and the thrush ducks into the vegetation and out of sight. I admit this tactic was annoying at first, but now I find it rather endearing.

The Swainson's thrush looks similar to another park bird, the hermit thrush, but the birds' tail feathers provide a useful identification clue. While the hermit thrush's bright, rust-colored tail contrasts with its brown back, the Swainson's brown tail blends in with its back of the same color.

NORTHERN WATERTHRUSH

(Parkesia noveboracensis)

There's something calming about watching a waterthrush stroll along the water's edge, bobbing its tail in fluid motion. When I first arrive at the Long Pond, the bird takes notice and darts behind the bright green thicket. But I know if I wait, it will return. I sit on the walking bridge and watch barn swallows gracefully skim the pond to quench their thirst. Within a few seconds, that familiar bobbing catches my eye—it's the waterthrush, weaving in and out of the vegetation. I watch as it edges out toward the pond to resume its search for tiny crustaceans and mollusks.

Northern waterthrush,
Pier 1 Long Pond

Northern waterthrush, Pier 1 Vale Lawn

*Black-throated blue warbler,
Pier 6 Exploratory Marsh*

BLACK-THROATED BLUE WARBLER

(Setophaga caerulescens)

Something small and dark darts up to a tree in the water playground. I have a hunch—and a hope—that it's a male black-throated blue warbler. But it's hidden now, so I have to wait. A few steps away in the Exploratory Marsh, I'm happy to see the Wilson's warbler is still around. I sit on one of the wooden planks of the garden's paths to immerse myself in fall migration (and to appear less intimidating than a five-foot-five human). A redstart lands on a branch above me; I watch as it moves from one leaf to the next, scarfing down insects. Out of the corner of my eye, I see a phoebe catching flies above the marsh.

Water gurgles as the fountains of the adjacent water playground start to spray. My thoughts return to the dark mystery warbler; will it retreat from the splashes and fly to the marsh? I feign lack of interest, which sometimes works, and start to turn my attention back to the redstart. But a bird flies in. It lands on a large flat rock where a puddle has formed. My hunch was right—it's a black-throated blue warbler, now bathing on the rock right in front of me.

Black-throated blue warbler (female),
Pier 1 Magical Knoll

 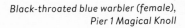

BLACK-THROATED GREEN WARBLER

(Setophaga virens)

On the first day of October 2015, the park was teeming with migrating birds. Over two dozen palm warblers were hopping around Harbor View Lawn, easily snatching insects that hovered over the blades of grass. Blackpoll warblers peeked out from the catalpas, their drab yellow fall plumage blending in with the trees' large leaves. I envisioned the predawn scene; as the flock was flying south, they spotted the skyscrapers of lower Manhattan on their right and the perfect stopover—Brooklyn Bridge Park—on their left. Touchdown.

Though the palms and blackpolls were plentiful, nothing stood out more than this solo female black-throated green warbler. Its contrasting plumage was beyond eye-catching as it foraged in the hawthorn trees that lined the lawn. Only males of this species sport the black throat; the throats of females and immature birds are white or pale yellow.

The simultaneous arrival and foraging proximity of the blackpolls, palms, and a black-throated green point to their being part of a mixed flock—birds of different species that move and forage together. Commonly seen during migration, mixed flocks are thought to be part of some birds' opportunistic strategy. Not only are there more eyes to search for food, there is strength in numbers when it comes to evading predators. Certain species even take advantage of the rejected or flushed prey of others.

Black-throated green warbler, Pier 1

*Yellow-bellied sapsucker,
Pier 1 Wine Bar*

YELLOW-BELLIED SAPSUCKER

(Sphyrapicus varius)

Many trees on Pier 1 (especially the oaks) have rows of small holes in their trunks. This is the drill work of the yellow-bellied sapsucker, a lively woodpecker that visits the park during migration. Once it has drilled holes, the bird sucks sap, of course, supplementing its sugar intake with insect protein that collects on the sticky stuff.

Sapsucker holes are called sap wells and are often used as a food source by insects, as well as hummingbirds and bats. One day, I came upon a strange sight—over half a dozen insect species were clustered on a tree at the Little Shrub Stand. As I moved closer, I saw an eastern comma butterfly (*Polygonia comma*) that had its proboscis in a sap well. Two Japanese beetles (*Popillia japonica*) also appeared to be imbibing in the sap cocktail. A scuffle ensued as other insects started to fight for a spot. I was so absorbed in this insect drama that I felt dizzy when I left.

The yellow-bellied sapsucker is the only North American woodpecker that is highly migratory. Birds that stop by the park in fall are likely headed to the southeastern United States or the West Indies.

SCARLET TANAGER

(Piranga olivacea)

One evening just before sunset, the Little Shrub Stand was as busy as the surrounding city streets. An ovenbird strutted past a giant rat. Two Carolina wrens flitted around the holly. A house wren peeked around a branch. And this scarlet tanager plucked berries from the dogwood while perched in an adjacent magnolia tree.

In its breeding plumage, the male scarlet tanager has a brilliant red body and black wings. In fall, it looks similar to the female, with a yellow-green body but darker, black wings. These birds breed in the eastern half of North America and winter in northern South America.

Scarlet tanager, Pier 1 Little Shrub Stand

Northern parula, Pier 1 Magical Knoll

NORTHERN PARULA

(Setophaga americana)

Standing on top of Granite Prospect, I direct my gaze north over Vale Lawn. The giant, heart-shaped leaves of a catalpa drape the periphery like a stage curtain as I watch the show. It starts with an out-of-this-world view of the Brooklyn Bridge, its arches flanked by moving ships and a still skyline. A character enters from a tree on stage left; it's the northern parula—a warbler unmistakable in its costume of blue-gray, yellow, and white. The small songbird flies quickly over people picnicking on the lawn below, then lands somewhere deep in the shade of a sumac.

There is nothing left to do at this point except enter the scene myself. I walk down to Vale Lawn and sit facing the Magical Knoll. And I wait. The East River Ferry sounds its departure horn—one long honk followed by three short ones—and races loudly past. Startled house sparrows and starlings—only extras in this show—flock to the catalpas for cover. Finally, the parula lands in a tree at center stage and I snap this photo from my front-row seat.

PINE WARBLER

(Setophaga pinus)

One of the benefits of birding in a park still in its infancy is the presence of short, young trees. This prevents quite a bit of neck strain and allows for close observation of species more difficult to observe in older parks with more mature, taller trees. As its name suggests, the pine warbler forages in pines, and often high in them. I found this one in its expected habitat—high in a pine along the Pier 4 path—yet this tree was only five feet high. I enjoyed my eye-level view as the bird checked cones and needle clusters for insects and larvae. It then flew across the path and landed on a low wire fence. As it moved along the wire, quickly placing one foot over the other, it reminded me of a yellow-costume-clad performer dancing on a tightrope.

The pine warbler arrives in the park later in fall (and earlier in spring) than most warblers, some of which start stopping by on their way south as early as July. The pine warbler's fall schedule—with a usual September arrival—prolongs the migration magic. By the time most migrants have already made their park pit stop, this warbler can be seen lighting up the trees as late as November.

Pine warbler, Pier 4

Field sparrow, Pier 3 uplands

FIELD SPARROW

(Spizella pusilla)

The field sparrow looks as if it has just returned from a visit to a Manhattan makeup counter. The delicate shade of salmon pink on its bill rivals the most complementary hue of lipstick, warming the grays and reddish browns on its face and crown. The tiny white feathers of its eye-ring make its eyes pop more than any eyeliner ever could.

In spring, the uplands of Pier 3 had been scant on good birds, the best being one very vocal northern mockingbird. Whenever I passed through, the only other species I saw were house sparrows. But one day a field sparrow darted out from some grass and landed in a tree. The mockingbird, intent on preserving its domination of the uplands, proceeded to chase the field sparrow clear to the other end of the lawn. This scenario repeated like clockwork; wherever the sparrow landed, the mockingbird followed. Before long, the chase traced a field-sparrow-pursued-by-mockingbird zigzag pattern over all points of the oval-shaped lawn.

A week later, on a busy Saturday, the mockingbird was perched on a tree overlooking the park goers that had taken over its turf. As it belted out an impressive repertoire of mimicked vocalizations, I spotted the field sparrow in the dried clumps of grass along the path. Now, with the mockingbird distracted, the sparrow was free to forage. I sat on the edge of a bench and watched the bird from a few feet away. It flew to a low branch and modeled its warm-hued coat and beauteous bill for the camera.

LINCOLN'S SPARROW

(Melospiza lincolnii)

There comes a point in many a birder's life where one bites the bullet and decides to study sparrow identification—the art and science of telling one small brown bird from another. At first, I found this prospect daunting. I couldn't imagine being able to confidently distinguish native New World sparrows, save for the song sparrow, a year-round park resident, and the white-throated sparrow of winter. But I felt my birding skills had reached a plateau, especially since I was birding the same location every day.

One fall, when a variety of sparrows began to dot the park's larger lawns, I knew it was time to step up my sparrow-identification game. I scoured field guides and studied photos. It was fun. Within a few weeks I was ready to try sorting out shades of brown and subtle behavioral differences in the field.

I had seen a Lincoln's sparrow once before but hadn't identified it myself and, new to birding, wasn't at all sure what I was seeing. But when a sparrow hopped out from the brush onto a metal grate walkway on Pier 1, I was ready. I noted thin streaking on the bird's breast set against a cinnamon color that branched out to the bird's flanks. These were all field marks of a Lincoln's sparrow. I reveled in my first confident identification of the species.

It took a year to spot another Lincoln's sparrow in the park, but it was worth the wait. This one perched in a shrub at the Long Pond and posed for this photo.

*Lincoln's sparrow,
Pier 1 Long Pond*

Golden-crowned kinglet, Pier 1 Harbor View Lawn

GOLDEN-CROWNED KINGLET

(Regulus satrapa)

One fall, dozens of golden-crowned kinglets, usually known for feeding high in trees, took to the lawns of Pier 1 in a foraging frenzy. Docile to the extreme, validating a reputation for tameness, they surrounded me as I sat on Vale Lawn. I held my breath as one hopped up and landed on my leg. Our eyes met, and within an instant the kinglet hopped back to the grass. I guess tasty insects were much more interesting than a big biped with binoculars.

This energetic little bird—one of the smallest of all North American song-birds—can survive temperatures below −40°F. When agitated, the male raises its yellow crown feathers to reveal an even more striking streak of reddish orange.

Golden-crowned kinglet shows red crown feathers, Pier 1

AMERICAN KESTREL

(Falco sparverius)

t was November 15, 2013, and my camera was only a few days new. While my first attempt at bird photos a day earlier had produced some nice shots of a cooperative hermit thrush and many robins, I still felt the camera was in control of me instead of the other way around. There was still much to learn, including how to comfortably hold it (was this even possible with such a big lens?) and how to walk without constant fear of the thing falling to the ground.

I made it to Pier 1 with camera still intact, expecting to encounter the same birds as I had the day before. As Harbor View Lawn came into view, there wasn't a person in sight. I walked north to check for thrushes on the grass along the steps. Before I rounded the corner of the lawn, I was welcomed by this American kestrel, perched on a light pole and staring straight at me. My reaction was unexpected; instead of delight, fear.

Would I get a good photo? How could this be happening on my second day of learning bird photography? Such thoughts filled my head as I awkwardly raised the camera.

In the end I think I did fairly well, considering how high this kestrel was perched. In fact, the next time I met this bird in the park, exactly one year later to the day, it was perched just as high and the photos were about the same. Looking back, it had been a lucky second day of bird photography.

Averaging under a foot in length, the American kestrel is the smallest falcon in North America. It nests in the eaves of buildings in nearby Brooklyn Heights and other surrounding neighborhoods, occasionally visiting the park to hunt for small rodents, birds, and large insects. The slate-blue wings indicate that it's a male; females' wings are reddish-brown.

American kestrel, Pier 1

Wilson's warbler, Pier 6 Exploratory Marsh

WILSON'S WARBLER

(Cardellina pusilla)

t was the first week of fall and Pier 6 seemed strangely still. Just a few days earlier, it had been alive with the sound of children laughing and splashing in the pier's many playgrounds. Another summer had come and gone. Now, the play areas served as bug buffets for migrating birds. A black-and-white warbler picked insects from holes in a metal grate below a wooden platform. A few feet away, a scarlet tanager dropped down to a small puddle for a drink. This Wilson's warbler foraged in the surrounding catalpa trees and eventually took a break, perching on a branch at the entrance to the slide playground.

Male Wilson's warblers are easy to identify by their small black cap; in females and immature birds the cap is very faint or not there at all. The bird's common name honors the Scottish naturalist Alexander Wilson. Often called the father of American ornithology, Wilson wrote and illustrated the nine-volume *American Ornithology*, published in 1808–1814. Here Wilson illustrated 268 species—26 of them new to science. There are more birds named after Wilson than any other ornithologist, including John James Audubon, his contemporary.

RUBY-THROATED HUMMINGBIRD

(Archilochus colubris)

E ven with dozens of park visitors milling about Pier 1, the Magical Knoll was alive with bird activity one day in May. I recorded thirty-five species in the park (a very high count for this location), including nine warbler species. Some of these were common sightings here during migration: ovenbird, northern waterthrush, black-and-white, common yellowthroat. But there were two warbler surprises—a black-poll and a female black-throated blue.

Hoping for a photo of the black-throated blue, I sat under a paulownia tree that over-looked the Magical Knoll. The tree's trumpet-like purple blooms filled the air with a sweet scent so powerful it overwhelmed the fumes of the nearby freeway and sightseeing choppers. In the distance, the arches of the Brooklyn Bridge peeked through the foliage. Warblers and flycatchers continued to zip around on the slope below. And then, just as I was getting comfortable, I heard a loud *bzzzzz* above. Thinking I might have disturbed a giant bumblebee, I carefully tilted my head back to look. Something dark and fairly large was moving in and out of the paulownia flowers. For a split second, I felt slight fright. Then I gasped as I realized what it actually was—this ruby-throated hummingbird.

The ruby-throated hummingbird is the only breeding (and regularly sighted) hum-mingbird in the eastern United States. The color of a hummingbird's throat feathers—collectively called a gorget—is produced by refracted light rather than pigments. In dim light, the ruby-throated's gorget can appear black. Had this bird appeared in direct sunlight and at the proper angle, its throat would have lit up like a gem.

Ruby-throated hummingbird,
top of Granite Prospect

Veery, Pier 1

VEERY

(Catharus fuscescens)

Whenever a thrush hopped out from the understory of Pier 1's wooded paths, I would raise my binoculars and wish that my eyes would fall upon the warm cinnamon-orange tones of a veery, a bird that was missing from my life (and park) list. But a turn of the focus knob always revealed a hermit thrush, another member of the genus *Catharus* that was slightly less secretive and had a much longer stay in the city. Don't get me wrong, I loved the hermit thrush, but at times I wondered if I would ever see a veery (a common feeling with a bird not yet on one's life list).

One day the trusty ol' Magical Knoll granted my wish, as a flash of something the color of a dried apricot caught my eye and disappeared deep into the shrubs at the knoll's base. It had to be a veery; no other Brooklyn bird had that unreal light amber color that I'd marveled at in countless Internet photos. The gorgeous thrush poked its head out from a shrub to snatch a berry. Later in the day, the little gem was hopping around the path lining Harbor View Lawn, where I got close enough for this photo.

TENNESSEE WARBLER

(Oreothlypis peregrina)

During migration, I'm always tempted to head straight for the park's birding hot spots: the Exploratory Marsh, the Magical Knoll, the Dark Forest. As much as I want to scour each and every inch of habitat in the park, there are only so many hours of light in the day. So I scan as much as I can while mostly just walking by certain mini-patches.

At the foot of a residential building on the north end of the park, there was a tiny stand of shrubs, about ten feet square, with a set of bike racks to its side. I had checked it a few times but never found much more than house sparrows and starlings. One day when I was beelining north to Pier 1, I heard the sound of a yellow warbler from behind the bike racks. Not only was the bird there to greet me, but I also found this Tennessee warbler gleaning insects from the goldenrod.

Tennessee warbler,
One Brooklyn Bridge Park

Brown creeper, Pier 1 Vale Lawn

BROWN CREEPER

(Certhia americana)

A tiny surprise nearly brings a tear to my eye on Vale Lawn—a brown creeper scaling an oak. I wonder how a bird so small and delicate can find its way through the maze of Manhattan to end up in my patch. Maybe it avoided the big island altogether and followed the East River. However it got here, I am stunned . . . and grateful to be spending time with this adorable bird.

The brown creeper isn't technically rare, but I only spot it in the park once or twice during migration. It can be easy to miss because its brown upper parts often blend in seamlessly with the bark on which it creeps. This sole North American member of the treecreeper family (Certhiidae) measures about five inches from tip of bill to end of tail. But this leads one to envision the bird as much bigger than it appears; its long bill and tail make up over half of its length.

Brown creepers usually scale a tree upward as they probe the bark for insects and insect eggs and pupae. And, once at the top, they tend to fly to another tree. I got lucky with this one; after scaling up the oak, it flew back to the bottom for another round of insect extraction.

ROSE-BREASTED GROSBEAK

(Pheucticus ludovicianus)

The rose-breasted grosbeak is an infrequent park visitor, but when it shows up, it's always on a good birding day. I had been out since sunrise, enjoying sightings of warblers on Pier 1—a black-throated blue hopped among the dawn redwoods near the Magical Knoll, an ovenbird strutted along the paths of the Dark Forest, and a northern parula foraged high in the catalpas on Granite Prospect. There was even a new warbler for my park list awaiting me at the Pier 6 Exploratory Marsh—the Nashville warbler.

When I went to check the Long Pond around midday, two skittish Swainson's thrushes fled from the willows. But one stout-looking bird with distinct head markings remained. I recalled seeing this bird once before, also on Pier 1. It was a female rose-breasted grosbeak—or so I thought. My photos proved me wrong; they showed a barely visible streak of red on the bird's breast and some reddish tint under its wing. This indicated it was a juvenile male. Only males sport the rose breast, and those feathers were just starting to grow in on this young bird.

This was now a very good birding day—I'd made a mistake and learned something new. I couldn't wait to search for this subtle distinction the next time I met a rose-breasted grosbeak.

*Rose-breasted grosbeak
(juvenile male), Pier 1 Long Pond*

Mourning warbler, Granite Prospect

MOURNING WARBLER

(Geothlypis philadelphia)

t's early fall. A small bird moves in and out of the shadows in the shrubs lining Granite Prospect. It looks a bit like a common yellowthroat, but different—darker, with a longer body—and it's much more secretive. I sit on the steps and watch it for a few moments before it disappears into the brush. Have I seen this bird before? I'm not sure, but I think it's a mourning warbler, though different from the one I had seen in spring. Adult males of this species have a gray hood, a feature to which the bird owes its name. But immature mourning warblers are drab, with less distinct markings, as are females in fall plumage.

Now comes the fun part; the sleuthing begins. I thumb through my many field guides and look for clues on the Internet. What is this bird? Other possibilities include a female common yellowthroat or an orange-crowned warbler. After my search, I still settle on an identification of mourning warbler but need confirmation. I email a photo to a few of my birding friends, and they all agree. Here we have a juvenile mourning warbler—and a rare sighting for the area.

LARK SPARROW

(Chondestes grammacus)

Just a few days into fall, I spotted this lark sparrow on a path at the Pier 3 uplands. It quickly disappeared into endless clumps of prairie grass (a perfect habitat for this sparrow), and I never saw it again. This rare sighting caused quite a stir among Brooklyn birders.

After I reported the sighting on eBird and the NYSbirds-L (the email discussion list for birds and birding in New York State), people arrived in hopes of finding it. It was great to see so many people on Pier 3 with binoculars around their necks. A few birders did spot the lark sparrow on the berm at Pier 3 uplands, but, unfortunately, some left empty-handed.

October and November are the best months to spot sparrows—including rarities—in the park and around the city. In November 2014, I had another rare sighting. A grasshopper sparrow hopped along the steps of Harbor View Lawn but headed straight for dense brush before I could snap a decent shot.

Lark sparrow, Pier 3 uplands

Sora, Pier 1 Vale Lawn

SORA

(Porzana carolina)

When birding, it helps to know what species to expect at your location and when (using eBird is a great way to find out). But it's also important to expect the unexpected, or, in the words of inspiring birder David Lindo (aka "The Urban Birder"), to "think that anything can turn up anywhere at anytime." On my usual birding rounds, I went to check Vale Lawn and the adjacent Magical Knoll. There, in the middle of the lawn, was a sora. I quickly snapped some photos from a distance; I knew that once I moved any closer, this secretive bird would sprint out of sight. And it did just that when I approached, flying deep into the brush of the Magical Knoll. (This hot spot was certainly living up to its name.) After reporting the sora—a rare sighting for the area—birders began to arrive. Fortunately, the bird moved to the Long Pond where it could be easily observed.

Though the sora is considered a rare sighting in Brooklyn Bridge Park, it is the most common rail (member of the family Rallidae) in North America. Even where common, the bird can be difficult to spot; it often remains hidden in the marsh grasses of its preferred wetland habitat.

TUNDRA SWAN

(Cygnus columbianus)

As I approached Pier 5, I noticed the floating platforms for the soon-to-be marina had been delivered. A giant crane barge was docked at the pier that would move them into place. I wondered—would the new marina deter the winter diving ducks from swimming close to shore? Then, something unusual caught my eye: a swan, swimming inside the perimeter of the platforms. I had only seen swans flying past the park—mute swans—but none had ever stopped by for a visit. I wasn't sure what type of swan this was, though. It had an interesting bright pink bill that also showed some black. I watched as it foraged in the water alongside mallards; then, for some reason, it started following a certain Canada goose, swimming close behind it near Bird Island.

When I got home I consulted every single one of my field guides and visited some Web sites, but I only found one match. The swan I saw at Pier 4 looked exactly like the illustration of the juvenile tundra swan in *The Sibley Guide to Birds.* Just to make sure (identifying juvenile birds in their various stages of plumage and bill development can often be tricky), I posted a photo on Twitter and asked others if they thought it was a tundra swan. Other birders agreed. I had hit the jackpot—not only was the tundra swan rare for the area, it was also a lifer. And my 130th park species.

Tundra swans breed in the coastal plains of the Arctic. Eastern populations (those that winter in the eastern United States) usually bypass the city on their way to their preferred wintering locations—freshwater and brackish wetlands, primarily along the Atlantic Coast from Maryland to North Carolina. I wonder what attracted the swan to Pier 4. Maybe it was those new marina platforms. . . .

Tundra swan (juvenile), Pier 5

Red-tailed hawk, Pier 1

RED-TAILED HAWK

(Buteo jamaicensis)

I was on a mission one day in early November 2015: Find the American woodcock. My friend Peter and I had seen one in flight just days before, and I desperately wanted to get a photo of the bird. I scanned the leaf litter (a favorite hiding place for the "timberdoodle," as it's often called) behind the benches at Bridge View Lawn. It seemed strangely quiet for this point in fall—not a junco, song sparrow, or flicker was foraging. Then, in the distance, I saw a creature the size of a small horse standing on the grass. I curiously looked through my binoculars and saw the piercing eyes of this red-tailed hawk. (The hawks measure about two feet from bill to tail, but the setting and surprise made it seem larger than it really was.) It must have just gone in for a kill, its intended victim likely being a rat, pigeon, squirrel, or starling. The empty-taloned raptor flew up to a light post that looked out over Harbor View Lawn. At takeoff, it flew south, circling over the water with the Statue of Liberty in the background.

The most common hawk in North America, the red-tailed hawk is a generalist species that thrives in many different environments and on a wide variety of food sources. New York City has many resident red-taileds—the most famous being Central Park's Pale Male, whose story was told in Marie Winn's *Red-Tails in Love*. Winter offers the best chance to sight these birds of prey flying over the city because there are more of them; many northern red-tailed hawks migrate south to winter in the city.

Park Bird List

My quest to spot and photograph the birds of Brooklyn Bridge Park began on April 1, 2013. This book covers my adventure (still ongoing) through November 30, 2015. Below are the common names of all birds in the order I found them. Birds whose names are in bold smiled (or frowned) for the camera, and their photographs—all taken in Brooklyn Bridge Park—are included in this book.

1. **Gadwall**
2. **Mallard**
3. **Double-crested Cormorant**
4. **Ring-billed Gull**
5. **Rock Pigeon**
6. **Eastern Phoebe**
7. **American Robin**
8. **White-throated Sparrow**
9. **House Sparrow**
10. **Brant**
11. **American Black Duck**
12. **Northern Mockingbird**
13. **European Starling**
14. **Song Sparrow**
15. **Northern Cardinal**
16. **Canada Goose**
17. **Red-breasted Merganser**
18. **Mourning Dove**
19. **Palm Warbler**
20. Common Loon
21. **Hermit Thrush**
22. **Ruby-crowned Kinglet**
23. **Dark-eyed Junco**
24. **Fox Sparrow**
25. **Common Grackle**
26. **Laughing Gull**
27. **Northern Flicker**
28. **Eastern Towhee**
29. **Great Black-backed Gull**
30. **Brown Thrasher**
31. **Red-tailed Hawk**
32. **Herring Gull**
33. **Blue Jay**
34. **Barn Swallow**
35. Killdeer
36. **Blue-headed Vireo**

37. **Red-winged Blackbird**
38. **Gray Catbird**
39. **American Crow**
40. **Savannah Sparrow**
41. Spotted Sandpiper
42. **Common Tern**
43. **Green Heron**
44. Chimney Swift
45. **Cedar Waxwing**
46. **Brown-headed Cowbird**
47. **American Redstart**
48. **Magnolia Warbler**
49. **Baltimore Oriole**
50. House Wren
51. **Least Flycatcher**
52. **Swainson's Thrush**
53. **Northern Waterthrush**
54. **Common Yellowthroat**
55. **Black-throated Blue Warbler**
56. **Black-throated Green Warbler**
57. Least Sandpiper
58. **Yellow-bellied Sapsucker**
59. **Black-and-white Warbler**
60. **Swamp Sparrow**
61. **Scarlet Tanager**
62. **Northern Parula**
63. **Ovenbird**

64. **Pine Warbler**
65. Red-breasted Nuthatch
66. **Field Sparrow**
67. **Yellow-rumped Warbler**
68. **Chipping Sparrow**
69. **Lincoln's Sparrow**
70. Osprey
71. **Golden-crowned Kinglet**
72. Pied-billed Grebe
73. **American Kestrel**
74. American Goldfinch
75. **Bufflehead**
76. **Downy Woodpecker**
77. **Greater Scaup**
78. **Red-necked Grebe**
79. **Red-throated Loon**
80. Hooded Merganser
81. Blue-gray Gnatcatcher
82. **Wood Thrush**
83. **Yellow Warbler**
84. **Warbling Vireo**
85. **Eastern Kingbird**
86. **Chestnut-sided Warbler**
87. **Wilson's Warbler**
88. **Eastern Wood-Pewee**
89. Great Blue Heron
90. **Blackpoll Warbler**

91. Ruby-throated Hummingbird
92. Red-eyed Vireo
93. Black-crowned Night-Heron
94. Louisiana Waterthrush
95. Yellow-billed Cuckoo
96. Veery
97. Cape May Warbler
98. Peregrine Falcon
99. Great Crested Flycatcher
100. Philadelphia Vireo
101. Mute Swan
102. Tennessee Warbler
103. White-breasted Nuthatch
104. Nashville Warbler
105. Tufted Titmouse
106. Brown Creeper
107. Black-capped Chickadee
108. Red-bellied Woodpecker
109. Winter Wren
110. Grasshopper Sparrow
111. Northern Shoveler
112. Ruddy Duck
113. Canvasback
114. American Coot
115. Wood Duck
116. Marsh Wren

117. Yellow-throated Vireo
118. White-crowned Sparrow
119. Rose-breasted Grosbeak
120. Mourning Warbler
121. House Finch
122. Fish Crow
123. Great Egret
124. Indigo Bunting
125. Carolina Wren
126. Willow Flycatcher
127. Blue-winged Warbler
128. Lark Sparrow
129. Sora
130. Tundra Swan
131. American Woodcock
132. Bald Eagle
133. Purple Finch
134. American Tree Sparrow

There are two birds not on the list that do appear elsewhere in the book: an *Empidonax* flycatcher that I was unable to identify to species and an escaped exotic budgerigar parakeet.

Note: No food or birdcall audio was used to obtain the photographs in this book.

The Plight of Urban Birds

Birds face many struggles—finding food (both for themselves and their young), evading predators and brood parasites, competing for nest locations—the list is long. Migrating birds must deal with storms, habitat loss in stopover locations, and exhaustion. Urban birds face yet another danger—window collisions.

During fall migration in 2015, there was a northern waterthrush that foraged behind the wine bar at Pier 1. Every day, I would check to see if it was still there—and it was—for an entire month. When I didn't see the waterthrush for three days straight, I assumed it had finally headed south.

A few days later, I was notified via email of the location of a dead unidentified warbler just outside the park. I was, albeit with a sense of dread, hoping it would still be there the next day so I could document it and enter it into D-bird (d-bird.org), NYC Audubon's database for tracking birds that are victims of window strikes. When I arrived, the bird was still there—a dead, still strikingly beautiful northern

waterthrush. My heart sank as I snapped a photo of the bird for the records. And then it hit me—could this be the wine bar waterthrush? I walked just across the street to the wine bar; there was no sign of the bird. Desperately, I looked for the bird's bobbing tail, but the area was still and silent.

Birds deal surprisingly well with many threats and dangers, but they appear defenseless against the reflective properties of glass or, just as perilous, its transparency. Fortunately, NYC Audubon has spearheaded efforts to deal with the issue and works with organizations to install bird-safe glass on their buildings or take other ameliorative measures. Places that were once major strike locations—

including Manhattan's Jacob K. Javits Convention Center and Time Warner Center—have seen a great reduction in fallen birds. But the problem is far from solved. Reporting dead birds into the D-bird database helps NYC Audubon track problem areas around the city and helps them help urban birds.

There are efforts to address this issue on a national level as well. The Federal Bird-Safe Buildings Act, introduced in 2015, would require all newly constructed federal government buildings and those undergoing a substantial alteration to use bird-friendly building design, including the use of bird-safe glass. The act would also require minimal use of external lighting, which disorients birds and causes them to strike windows or circle in flight until they are so exhausted they drop from the sky. Another effort, National Audubon Society's Lights Out, encourages building owners and managers to turn off unnecessary lighting during bird migration periods. New York State recently joined this effort; in 2015, Governor Andrew Cuomo announced that all state-owned and state-managed buildings would turn off bright outdoor lights at night during specified migration periods. NYC Audubon is also very active in recruiting city organizations to participate in Lights Out.

Notes on the Camera-Shy Species

Below are notes on the birds spotted on my quest whose photos do not appear in the book. A few of the birds I saw before I bought my camera. Others were perched high, soaring in the sky, or foraging close but in extremely low light. Though many of these birds are commonly found elsewhere in New York City, most are not regular park visitors. For some, especially ducks that prefer freshwater lakes, the park simply does not offer their preferred habitat. Other birds may start to visit more often when the trees mature or ecological conditions change. And some might just not have discovered the park yet. The Carolina wren, who appears in the book, seems to have just recently discovered the park's wine bar and its surrounding habitat. Rarely seen prior to 2015, it is currently a daily sighting. Frequency and distribution of many birds change over time because of habitat creation or loss or other ecological factors.

The list below contains a few shorebirds—the only ones I've spotted in the park. Shorebirds are rarely seen here because most need plenty of beach, shallow water, and/or mudflats in which to feed. They are, however, a birding highlight of New York City summers, when thousands stop by places like Jamaica Bay Wildlife Refuge on the way to their wintering grounds in the southern United States and South America. Some also nest on city beaches. Some species are federally protected, their nesting sites off-limits to beach visitors.

Mute Swan (*Cygnus olor*): Mute swans are not actually mute. I heard a pair honking as they flew down the East River. They get their name from their relatively muffled call (calls of other swan species carry longer distances). A year-round resident in some New York City ponds, lakes, and bays, this introduced species is native to Europe and Asia.

Hooded Merganser (*Lophodytes cucullatus*): This diving duck prefers to winter in freshwater lakes and bays (and does in other city locations). It can also be found in brackish bays and estuaries such as the East River, where I once spotted it. The hooded merganser has a large, striking crest that it can raise and lower.

Ruddy Duck (*Oxyura jamaicensis*): A small duck with a short, stiff tail, the ruddy reminds me of a classic rubber duck. Though not yellow, it has a small blue bill that adds to its toylike appearance. One winter, I found a lone ruddy duck resting on the water between Pier 2 and 3. These ducks can be found during winter in the city. Some even stick around for the summer.

View from main park path, south of Pier 1

Common Loon (*Gavia immer*): I've spotted this loon a few times in winter but not as often as the red-throated. When it visits it can usually be seen diving between the ends of the park's piers, close to the East River. In their nonbreeding winter plumage, common and red-throated loons look similar, but the common is larger and stouter. It also tends to hold its bill level; the red-throated holds its bill a bit upturned.

Pied-billed Grebe (*Podilymbus podiceps*): I'll never forget the first time I saw a pied-billed grebe do its disappearing act in Florida. The minute I laid eyes on it, it sank like a mini submarine. These birds can control their buoyancy; they trap water in their feathers, which enables them to sink partially with heads still above water—or fully (when I show up).

Great Blue Heron (*Ardea herodias*): This is the one that got away. A great blue heron had been visiting Pier 1 in late summer 2015. Locals mentioned they had seen it wading by the salt marsh. Once, it ascended from the freshwater gardens right behind me. On another occasion, the large heron erupted from behind the catalpas and flew over me while I stood on Vale Lawn. I had high hopes to photograph the bird in the park, yet it appears to have ended its regular park visits.

The great blue is the largest heron in North America and stands about three feet tall, though when it stretches it can reach up to four. It's easily spotted at Jamaica Bay Wildlife Refuge in summer and fall.

Osprey (*Pandion haliaetus*): The osprey is a fish-eating raptor occasionally seen flying over the park. I've seen it a few times over the southern piers. These hawks nest at Jamaica Bay Wildlife Refuge, where they can be easily spotted throughout spring and summer. While an osprey nesting platform currently exists at Bird Island, it has mostly attracted starlings and house sparrows. Set low and close to shore, the platform may be too near to humans to attract the osprey. And the park is especially crowded in summer, just when the bird would be using the nest.

Bald Eagle (*Haliaeetus leucocephalus*): In fall 2015, a man stopped to ask me about birds in the park. Before he left, he said, "Oh, and one more thing—ever see any bald eagles here?" I told him there were several locations in the city where they nested, but that it was unlikely to spot a bald eagle in the park. The very next day, while standing at the Long Pond, I saw something soaring high. I quickly raised my binoculars and zoomed in on a juvenile bald eagle. Never say never.

American Coot (*Fulica americana*): Many nonbirders mistake the American coot for a duck—and it's easy to do. Coots are found in open water and often swim right alongside ducks. But the coot is a rail (member of the family Rallidae), most of which are secretive birds that stay hidden within marsh grasses. I once saw a lone coot swimming near the Pier 1 pilings, but it stayed only a day.

During winter, coots are commonly found in city wetlands and lakes.

Killdeer (*Charadrius vociferus*): A shorebird that prefers large open spaces, including parking lots, the killdeer is known for its broken-wing display. When a predator or perceived threat comes close to its nest, the killdeer moves off a short distance, feigning injury and luring the intruder farther and farther away.

There used to be a killdeer on the open lot where the Pier 4 beach and Bird Island were later built. Once construction for these features started, the bird moved on. It's ironic that the creation of Bird Island and a beach drove this shorebird away. Habitat management involves such a delicate balance. Yet there is a consolation. Every

Lights of Pier 6 glow at sunset

day, I hear the killdeer's call—*kill-deer, kill-deer, kill-deer*—sung in perfect mimicry by a park mockingbird.

Spotted Sandpiper (*Actitis macularius*): This medium-size shorebird took me by surprise when I saw it teetering on a beam under Pier 6. In summer plumage, when it shows bold and pretty spots on its belly, breast, and flanks, the spotted sandpiper is one of the easier shorebirds to identify.

Least Sandpiper (*Calidris minutilla*): I once saw three of these small shorebirds feeding on the rocks by Pier 5. In summer, the least sandpiper is a common sight at Jamaica Bay Wildlife Refuge, where it stops during migration.

American Woodcock (*Scolopax minor*): For over five years, the American woodcock was my nemesis bird—a species that all of my birding friends had seen but I had not. I even went on a woodcock watch to see the bird's nocturnal courtship flight. Not a single woodcock showed up. A secretive shorebird that actually avoids the shore, the woodcock is often found hiding in leaf litter during winter. It does show up regularly in the city, even in Bryant Park, just a couple of blocks from Times Square. Being a city birder, I should have seen the bird my first year back in Brooklyn.

In fall 2015, I met Peter in the park and told him flatly that today was the day we would see an American woodcock. We set out, catching up on the week's events and eyeing the dead leaves for a timberdoodle, another name for the woodcock. As I was admiring a gray catbird behind the wine bar, Peter said matter-of-factly, "I won't tell you what I just saw." Figuring he had seen an interesting sparrow, I followed him. Several minutes later, something flew off. "There it is," he said, as an American woodcock rounded the corner of the wine bar. I ran up to Peter and gave him a big hug. A magical birding moment.

Yellow-billed Cuckoo (*Coccyzus americanus*): The yellow-billed cuckoo is one of my favorite park sightings to date. With its long polka-dotted tail and bold yellow eye-ring, it looks much too tropical to be hanging out in the city. One fall it peered out from a dawn redwood at the base of the Magical Knoll.

The yellow-billed cuckoo is one of the few birds that can digest hairy caterpillars and has been observed eating over a hundred tent caterpillars in one sitting.

Chimney Swift (*Chaetura pelagica*): In May and June, chimney swifts can often be seen (and heard) from the park's northern piers as they flock over Brooklyn Heights. Occasionally, they also fly over the park. Swifts feed on flying insects and spend most of their time in the air.

Chimney swifts can't perch like most birds, because their feet are adapted to grasp vertical surfaces. As their name suggests, they often nest in chimneys; once Europeans colonized North America, swifts started using these handy manmade structures instead of hollow trees.

Red-bellied Woodpecker (*Melanerpes carolinus*): The red-bellied woodpecker can be found year-round in the city. This bird rarely visits Brooklyn Bridge Park, but that may change as the park's trees mature. The red-bellied prefers medium to large trees.

Red-bellieds may seem oddly named. The red belly is very faint. The bird's red head (in males) and nape (in males and females) are what stand out, but the red-headed woodpecker already had dibs on this name.

Peregrine Falcon (*Falco peregrinus*): The peregrine falcon is known for reaching speeds of over 200 mph during its hunting dives and can occasionally be seen flying through the park. I once enjoyed a close view of one flying south between Piers 4 and 5. Since 1999, they've been nesting just across the East River on a window ledge of a Manhattan office building at 55 Water Street. A perch and nest area was built on the ledge, which likely keeps the falcons coming back. And locals often tell me of the "good ol' days," when a pair used to nest on the Brooklyn Bridge.

Great Crested Flycatcher (*Myiarchus crinitus*): I spotted this large, colorful flycatcher—with its yellow belly, reddish-brown upper parts, and gray head—one fall in a dawn redwood on Vale Lawn. This hole-nester likes adorning the nest with a shed snakeskin.

Philadelphia Vireo (*Vireo philadelphicus*): This was my 100th park species and one of the few camera-shy birds I observed at close range. If it weren't for the catalpa trees providing such perfect shade, I might have captured a properly exposed photo. But then again, if the large catalpa leaves hadn't been there, the bird probably wouldn't have stopped. The Philadelphia vireo is rare for New York City, though a handful usually show up during migration.

Black-capped Chickadee (*Poecile atricapillus*): The tiny black-capped chickadee has a recognizable buzzy song. One day, I heard it on Pier 1—*chickadee-dee-dee*. I followed the sound to find the chickadee foraging high in a plane tree. This is a common city bird that visits parks with feeders, but I rarely see it in Brooklyn Bridge Park.

Tufted Titmouse (*Baeolophus bicolor*): A year-round city bird, the tufted titmouse is a small, energetic songbird with a prominent crest. It's also a big fan of bird feeders and rarely visits the park, but this is another species that we might see a bit more of as trees mature.

Red-breasted Nuthatch (*Sitta canadensis*): The red-breasted nuthatch breeds in conifer

Spiral Pool, Pier 2

forests in the northeastern and western United States and Canada, where it often stays year-round. In winter, the bird's diet consists largely of conifer seeds. In years when cones are scarce, the bird irrupts; large numbers of nuthatches move out of their normal range, heading farther south in search of food.

White-breasted Nuthatch (*Sitta carolinensis*): The white-breasted nuthatch is another common city bird that is attracted to feeders. Like other nuthatches, it forages on tree trunks and branches.

House Wren (*Troglodytes aedon*): The house wren is a regular visitor to the park during migration. I have seen it at several locations, including the Exploratory Marsh, the wooded paths of Pier 1, and the terrace of the Pier 3 uplands. But for some reason, the bird has proven difficult for me to photograph. The house wren has the widest distribution of any songbird in the New World; it can be found from Canada to the southernmost point of South America.

Blue-gray Gnatcatcher (*Polioptila caerulea*): The blue-gray gnatcatcher is a migrant that can be found in the city during its summer breeding season. It feeds on many types of insects, with gnats making up only a small portion of its diet.

Nashville Warbler (*Oreothlypis ruficapilla*): The Nashville warbler passes through the park during spring and fall migration. I saw one frequenting the Long Pond one fall; on another occasion, a Nashville was foraging in the Exploratory Marsh.

Cape May Warbler (*Setophaga tigrina*): The Cape May warbler visits the city during migration. It breeds in the boreal forest of Canada and the northern United States and winters in the West Indies. It is one of many birds first described to science by the Scottish naturalist Alexander Wilson; Wilson first spotted this warbler species in Cape May, New Jersey. Interestingly, after the warbler received its common name, it was not recorded in Cape May even once during the next hundred years.

Grasshopper Sparrow (*Ammodramus savannarum*): As I walked up the steps lining Harbor View Lawn, I noticed an interesting sparrow on a step way ahead. When I zoomed in with my lens, I was able to make it out—a grasshopper sparrow. This bird, rare for the area, not only eats grasshoppers, it also sounds like one.

Indigo Bunting (*Passerina cyanea*): Indigo buntings get their name from the male's deep blue breeding plumage. It's hard not to notice these striking males when they fly by. Though I've seen the indigo bunting only once in the park—by the Long Pond—it can be reasonably expected during migration.

American Goldfinch (*Spinus tristis*): These small songbirds occasionally visit the park in flocks. They can be found year-round in city parks with bird feeders. During breeding season, the male is quite striking with its bright yellow and black plumage.

Purple Finch (*Haemorhous purpureus*): Amid the squeaks and squeals of starlings, I heard something quieter and more repetitive as I walked the wooded paths. The soft tone was difficult to track, but after straining my ears (and my neck) I found my 133th species—a pair of purple finches foraging high atop an oak.

American Tree Sparrow (*Spizella arborea*): As I passed by the marsh on Pier 1, I noticed two American goldfinches in a shrub. They flew away quickly but made me stop and take a closer look at the marsh. I'm glad I did. Among many white-throated sparrows was a bird with a reddish-brown cap and dark breast spot—an American tree sparrow. I admired my 134th park species as it snacked on aster seeds while perched low and in the shadows.

Barn swallow fledglings, Pier 3

Birds Commonly Sighted in Brooklyn Bridge Park

Below are the species that are commonly found and easiest to spot in Brooklyn Bridge Park based on my sightings from April 1, 2013–November 30, 2015.

The goal of this section is to present the birds you are likely to see on a stroll through the park. It is not a comprehensive list. For that information, you can consult up-to-date, detailed bar charts of sightings by month (even by week) for Brooklyn Bridge Park—or your own location—at Cornell Lab of Ornithology's eBird (ebird.org).

In the seasonal lists below, birds are placed into the season during which they are most abundant and likely to be observed. For example, the gray catbird arrives in spring and departs in fall, but you are most likely to spot it in summer. Species that nest in the park are denoted by an asterisk.

Year-round Park Residents

Canada Goose

American Black Buck

Mallard*

Double-crested Cormorant

Ring-billed Gull

Herring Gull

Great Black-backed Gull

Rock Pigeon

Mourning Dove*

American Robin*

Northern Mockingbird*

European Starling*

Song Sparrow*

Northern Cardinal

House Sparrow*

The Carolina wren may have also recently become a year-round resident. I first spotted one in spring 2015 and now, deep into fall 2015, I am seeing one of these wrens daily by the wine bar.

Winter

Brant
Gadwall
Bufflehead
Red-breasted Merganser
White-throated Sparrow

Summer

The birds of the park summers are listed below. In late summer, fall migrants start to appear; the yellow warbler is one of the first to stop by on its way south. It's more often seen in the park in August and early September than in later fall months.

Laughing Gull
Common Tern
Barn Swallow*
Gray Catbird*
Common Grackle*
Yellow Warbler

Spring and Fall

In general, it's easier to see many of the following birds in fall—when migrants stay longer—than in spring—when they are in a hurry to get to their respective breeding grounds. Some of these birds—such as the downy woodpecker and eastern towhee—can be found year-round in other locations in New York City where they breed.

Eastern Phoebe
Downy Woodpecker
Northern Flicker
Ruby-crowned Kinglet
Hermit Thrush
Dark-eyed Junco
Swamp Sparrow
Eastern Towhee
Ovenbird
Northern Waterthrush
Black-and-white Warbler
Common Yellowthroat
American Redstart
Palm Warbler

Rare Sightings
in Brooklyn Bridge Park

A s of this writing, I have had the following rare sightings (of birds rare for the location and/or time of year) in Brooklyn Bridge Park.

Tundra Swan (October 19, 2015)

Sora (October 10, 2015)

Philadelphia Vireo
(September 16, 2014)

Grasshopper Sparrow
(November 13, 2014)

Lark Sparrow
(September 26, 2015)

Mourning Warbler
(May 25, 2015, and
September 17, 2015)

*View of Manhattan
from Pier 1 Salt Marsh*

Tips for Getting Started with Birding

I hope the many birds of Brooklyn Bridge Park have inspired you to take a look around your own neighborhood or local park. Here are a few tips for getting started with birding.

- **Get a field guide.** I started with *National Geographic Field Guide to the Birds of North America*, but there are many options. Get the one that appeals to you and inspires you to look for birds. Once you have one, take a look through and see if you recognize any of your neighborhood birds. Bring the guide with you and pick a bird to identify. A bird's body and bill shape can often give you more clues than plumage colors, especially when beginning birding.

- **Get some binoculars.** Standard, popular binocular sizes for birders are 8 × 42 and 10 × 42. The first number indicates magnification, the second indicates the objective lens size (diameter of the large end of the binocular in millimeters). The larger the objective lens size, the more light enters the lens and the brighter the image. I've used and like both sizes; 10 × 42 are more sensitive to hand shake. For kids and those who have trouble finding objects through binoculars, 6x magnification works well; a 6 × 30 pair is lightweight and provides a wide and deep field of view.

- **Know what to expect in your area.** Visit eBird (ebird.org) and choose the "Explore Data" option to discover birds that are showing up in your park as recently as the very day you are inquiring.

- **Attend a meeting of your local Audubon chapter.** Visit audubon.org to find a chapter near you.

- **Have fun.** Birding can be frustrating at times (like when there isn't a bird in sight). Knowing what to expect (see above) can make birding extremely fun. And after you've experienced a few magical birding moments, you may just get hooked on a lifetime hobby.

What about apps?

There are also many field guide apps available, which are a great supplement to a book format. Try Cornell Lab of Ornithology's Merlin Bird ID (perfect for beginners) or Audubon Bird Guide: North America.

A note on the use of birdcall audio: Many apps contain birdcall audio, a useful resource for learning to recognize bird sounds. There is an increasing trend to use these audio tracks to attract birds, which may have a negative effect on them.

In my early days of birding in Florida, I called out a groove-billed ani from a tree with an app audio track. It hopped down and looked around for a friend—a fellow member of its species. It might have even thought it had found a mate. I felt an extreme sense of guilt and stopped using audio. Studies have shown that in certain species, when a female hears a male sing, it triggers hormonal changes that ready her for mating. Several times during my quest, I was tempted. Finding a bird and getting a photo would be so much easier if I called it in. I'm glad I never did.

The use of birdcall audio is prohibited in certain national parks and other places. I liken its use to playing "ding-dong-ditch" with birds. Whatever you decide, please be respectful of birds in the field; if you must use audio it should be used sparingly and not during breeding season or when observing a rare bird.

Selected Bibliography

Alderfer, Jonathan, and Jon L. Dunn. *National Geographic Birding Essentials: All the Tools, Techniques, and Tips You Need to Begin and Become a Better Birder.* Washington, DC: National Geographic, 2007.

"All About Birds." Cornell Lab of Ornithology. allaboutbirds.org. Accessed October 3, 2015.

"The Audubon Dictionary for Birders." National Audubon Society. January 13, 2015. audubon.org/news/the-audubon-dictionary-birders. Accessed November 11, 2015.

Brinkley, Edward S. *National Wildlife Federation Field Guide to Birds of North America.* New York: Sterling, 2007.

"D-Bird: NYC Audubon Bird Mortality Database." NYC Audubon. d-bird.org. Accessed October 3, 2015.

Dunn, Jon L., and Jonathan Alderfer. *National Geographic Field Guide to the Birds of North America.* 6th ed. Washington, DC: National Geographic, 2011.

Dunne, Pete. *Pete Dunne's Essential Field Guide Companion.* Boston: Houghton Mifflin, 2006.

Earley, Chris G. *Warblers of the Great Lakes Region and Eastern North America.* Toronto: Firefly Books, 2003.

"eBird." Cornell Lab of Ornithology. ebird.org. Accessed October 3, 2015.

Ehrlich, Paul R., David S. Dobkin, and Darryl Wheye. *The Birder's Handbook: A Field Guide to the Natural History of North American Birds: Including All Species That Regularly Breed North of Mexico.* New York: Simon & Schuster, 1988.

"The Feather Atlas: Flight Feathers of North American Birds." US Fish and Wildlife Service Forensics Laboratory, 2010. fws.gov/lab/featheratlas/index.php. Accessed November 3, 2015.

Haupt, Lyanda Lynn. *The Urban Bestiary: Encountering the Everyday Wild.* New York: Little, Brown, 2013.

Kaufman, Kenn. *Kaufman Field Guide to Advanced Birding: Understanding What You See and Hear.* New York: Houghton Mifflin Harcourt, 2011.

McGowan, Kevin J., and Kimberley Corwin, eds. *The Second Atlas of Breeding Birds in New York State.* Ithaca, NY: Comstock Publishing Associates, 2008.

Roberts, Dina L. "Conservation Value of Forest Fragments for Wood Thrushes (*Hylocichla Mustelina*) in Costa Rica's Caribbean Lowlands." *Revista Latinoamericana de Conservación | Latin American Journal of Conservation* 2, no. 1 (2011).

Rodewald, Paul, ed. The Birds of North America Online: bna.birds.cornell.edu/BNA. Cornell Laboratory of Ornithology, Ithaca, NY, 2015.

Sibley, David. *The Sibley Guide to Birds.* New York: Alfred A. Knopf, 2000.

Stephenson, Tom, and Scott Whittle. *The Warbler Guide*. Princeton: Princeton University Press, 2013.

Weidensaul, Scott. *Of a Feather: A Brief History of American Birding*. Orlando, FL: Harcourt, 2007.

Wilkins, Khristi A., Richard A. Malecki, Patrick J. Sullivan, Joseph C. Fuller, John P. Dunn, Larry J. Hindman, Gary R. Costanzo, Scott A. Petrie, and Dennis Luszcz. "Population Structure of Tundra Swans Wintering in Eastern North America." *Journal of Wildlife Management* 74, no. 5 (July 2010): 1107–11.

Winn, Marie. *Red-Tails in Love: A Wildlife Drama in Central Park*. New York: Pantheon, 1998.

Young, Jon. *What the Robin Knows: How Birds Reveal the Secrets of the Natural World*. Boston: Houghton Mifflin Harcourt, 2012.

Acknowledgments

Many thanks to those who have supported and contributed to this project and book:

To Peter Joost, for editing the manuscript and sharing his immense knowledge of New York City birds.

To Ruby Chen, Meryl Greenblatt, Abe Hsuan, Bob Nirkind, Harold Moeller, Greg Wolf, and Vicki Wolf, for reading the initial proposal for this book and providing invaluable feedback that helped make it a reality.

To Brenda Callaway, for encouraging me to move forward with bird photography, for sharing her knowledge and talent, and for being a big inspiration, along with her husband Jerry Callaway, during my early days of birding and to this day.

To Charlie Chessler and Janet Zinn, for their helpful bird photography tips and encouragement.

To Michael Yuan, for his enthusiasm about the birds of the park and for help with bird species verification and identification.

To Cindy Goulder and Matthew Wills, also for their enthusiasm about the park and for sharing their wide knowledge of plant and insect species.

To Doug Gochfield and Gabriel Willow, for taking the time to answer my last-minute questions.

To Cornell Lab of Ornithology, especially everyone on the eBird team, for all they do to help birds and to facilitate conservation efforts.

To Brooklyn Bridge Park Conservancy, including Eliza Phillips, Kara Gilmour, and Karla Osorio-Perez. And to the park's Director of Horticulture Rebecca McMackin and the entire gardening staff and volunteers, for creating and maintaining stunning park habitats that attract birds and insect pollinators.

To Brooklyn Bridge Park Corporation and Michael Van Valkenburgh Associates, for creating a beautiful urban green space with birds in mind, resulting in the park quickly becoming an important stopover for migratory birds.

To my mentors Bob and Lucy Duncan, for taking Connor and me under their wings and teaching us so much about birding and about life.

To David Lindo, for taking time out of his busy schedule to bird with me in my patch, and whose positive attitude and excitement for birding is beyond inspiring.

To Bob Wieler, for his endless support and for bringing such good birding vibes to our Sunday walks in Brooklyn Bridge Park, during which we had some of the most unbelievable sightings.

To everyone at The Experiment, especially Matthew Lore, my editor Nicholas Cizek, and designer Sarah Smith, as well as Jeanne Tao and Pamela Schechter.

To the people from all corners of the globe whom I've met while birding in Brooklyn Bridge Park—you are all part of my story. Thank you for inspiring me to continue my quest.

To the birds of Brooklyn Bridge Park and the world—you make life worth living.

And to the love of my life, John Connor, who took that birding trip with me during which we spotted our very first species—the belted kingfisher—and entered it into our little spiral-bound notebook. Thank you for your endless patience and support during the creation of this book.

Bird Index

Page numbers in italics indicate photographs

About the Author

HEATHER WOLF's love of birds was sparked while living on Florida's Gulf Coast, where she participated in the Florida Master Naturalist program and led walks for the Florida Trail Association. She currently lives in Brooklyn and works as a web developer for Cornell Lab of Ornithology and its eBird project. Her blog, brooklynbridgebirds.com, documents the birds of Brooklyn Bridge Park, where she has recorded over 30,000 bird sightings of more than 130 species. A woman of many passions, Heather has taught thousands of people to juggle and won the 2015 NYC Intel IoT (Internet of Things) Hackathon for her design of a "smart" juggling ball.